OPTO-ELECTRONICS

OPTO-ELECTRONICS

Lectures at the 1982 International Summer School
at The Chinese University of Hong Kong
26 July-6 August 1982

Charles K. Kao
P. K. Tien · Erich P. Ippen
Michiharu Nakamura

Edited by K. Young

The Chinese University Press
Hong Kong

International Standard Book Number: 962-201-283-3

The Chinese University Press
The Chinese University of Hong Kong
SHATIN, N.T., HONG KONG

Printing by Speed Printing Co., Ltd.

ADVISORY COMMITTEE

C.F. CHEN	The Chinese University of Hong Kong
B. HSU	The Chinese University of Hong Kong
Y. SUEMATSU	Tokyo Institute of Technology
L.B. THROWER	The Chinese University of Hong Kong
P.K. TIEN	Bell Laboratories
TSIEN San-tsiang	Academia Sinica
T. UCHIDA	Nippon Electric Company
C.N. YANG	State University of New York at Stony Brook
ZHOU Hua-sheng	Bureau of Science and Technology, Chinese Ministry of Posts and Telecommunication

ORGANIZING COMMITTEE

Chairman

P.K. TIEN	Bell Laboratories

Overseas Coordinator

Y.P. YAO	University of Michigan

Members

Y.W. CHAN	The Chinese University of Hong Kong
Y.W. LAM	The Chinese University of Hong Kong
H.T. NIEH	State University of New York at Stony Brook
D.C. XIAN	Academia Sinica

Secretary

K. YOUNG	The Chinese University of Hong Kong

PARTICIPANTS

ABDUL SALAM, Husin	Institute of Technology, Surabaya, Indonesia
BREEN, Daniel	KMIT, Bangkok
CHAN Sau Wah	University College, London
CHAN Yau Wah	The Chinese University of Hong Kong
CHAN Yiu Nam	The Chinese University of Hong Kong
CHEN Guofu	Xian Institute of Optics
CHEN Xihao	Chengdu Telecommunications Cables Research Institute
CHENG, Y.C.	University of Hong Kong
CHEUNG Nai Ho	Baptist College, Hong Kong
CHOY Chung Loong	The Chinese University of Hong Kong
CHUNG Suk Hing	Hong Kong Polytechnic
DAS, A.K.	Jadavpur University
DAHELE, J.	The Chinese University of Hong Kong
FUNG Yun Tong	The Chinese University of Hong Kong
GAO Fuyan	Shanghai Institute of Optics
GONG Xiaocheng	Jiao Tong University, Shanghai
HOI Chi Sam	The Chinese University of Hong Kong
HSU Dahsiung	Beijing Institute of Posts and Telecommunications
HUI Yuk Tak	The Chinese University of Hong Kong
LAM Yat Wah	The Chinese University of Hong Kong
LAU Kai Shui	Hong Kong Polytechnic
LAU Shing Tat	The Chinese University of Hong Kong
LAU Shun Yin	Hong Kong Polytechnic
LEE Kai Fong	The Chinese University of Hong Kong
LEE Wah Kwan	The Chinese University of Hong Kong
LEUNG, Thomas	Central University, Chung Li
LEUNG Wai Bor	Hong Kong Polytechnic
LEUNG Wing Pun	The Chinese University of Hong Kong
LI Qingxing	Zhongshan University, Guangzhou
LIM Pui Kong	Baptist College, Hong Kong

LIN Chongjie	Chengdu Telecommunications Engineering Institute
LIU Wenqing	Anhu Institute of Optics
LUA Kim Teng	National University, Singapore
MAK Siu Sang	Hong Kong Polytechnic
MONTUNO, M.	Ateneo de Manila University
NG Chiu King	The Chinese University of Hong Kong
POON Ming Cheong	The Chinese University of Hong Kong
POON Ying Ming	Hong Kong Polytechnic
SHUI Hailong	Shanghai Institute of Metallurgy
SO, Patrick	Hong Kong Telephone Co
TACHAGUMPUCH, A.	Chulalongkorn University, Bangkok
TAKAHASHI, N.	Keio University, Yokohama
TSUI Wan Lam	Hong Kong Polytechnic
WANG Shutang	Semiconductor Institute, Academia Sinica
WANG Wen Cheng	Fudan University, Shanghai
WONG King Young	University of Pennsylvania
WONG Sai Peng	The Chinese University of Hong Kong
YIP Chi Keung	The Chinese University of Hong Kong
YONG Yeow Chin	Mara Institute of Technology, Penang
YOUNG, Kenneth	The Chinese University of Hong Kong
YU Wing On	Baptist College, Hong Kong
ZHAO Luguang	Chengdu Institute of Technology
ZHAO Zisen	Wuhan Research Institute of Posts and Telecommunications

OPENING ADDRESS

Professor Baysung Hsu
Acting Vice-Chancellor
The Chinese University of Hong Kong

It gives me great pleasure to welcome you all to this International Summer School on Opto-electronics organized by The Chinese University of Hong Kong.

This University has always regarded as one of its roles the cross-fertilization of the cultures of East and West. The transfer of technological innovation to the developing regions of Asia is a matter of urgency to which we wish to make a modest contribution. In this connection, we have received advice from many distinguished scientists. It was out of discussions with these people that the idea of the Summer Schools was conceived.

Opto-electronics is a rapidly growing field, with potentially a great impact on what is regarded as the new age of information. We are very fortunate to have as lecturers in this programme four scientists who are prominent in this field.

This School is directed by Dr. P.K. Tien, Head of Electron Physics Research Department at Bell Laboratories. The other lecturers are Dr. Charles Kao, Vice-President and Director of Engineering of ITT Electro-optical Division; Dr. Erich Ippen, Professor of Electrical Engineering at M.I.T. and Dr. Michiharu Nakamura, Senior Researcher at the Central Research Laboratory of Hitachi Ltd. I am sure these lecturers are already known to you by reputation. I am particularly happy to see my friend and former colleague, Dr. Kao, who was the Chairman of the

Department of Electronics at this University from 1970 to 1974.

These lecturers are all very busy people and we are very grateful that they have been so generous with their time and effort.

This Summer School is financed in very large part by the Croucher Foundation of Hong Kong. Cathay Pacific Airways has provided reduced fares. Many participants are sponsored by IBM, UNESCO and COSTED (The Committee on Science and Technology in Developing Countries). May I take this opportunity to thank all these organizations for their support.

We are also grateful to the Hong Kong Telephone Company, which is co-operating with us in this endeavour by providing demonstrations of optical fibre installation at its facilities.

I am sure that you will gain a great deal from the four series of lectures and from discussions with one another. I wish you a very fruitful and enjoyable fortnight at this Summer School.

TABLE OF CONTENTS

Optical Communication Systems
Charles K. Kao

Integrated Optics and Integration of Optics with Electronics
P. K. Tien

Quantum Electronics, Picosecond Laser Pulses and Applications
Erich P. Ippen

Semiconductor Light Sources and Detectors
Michiharu Nakamura

OPTICAL COMMUNICATION SYSTEMS*

Charles K. Kao
ITT, New York, NY 10012, USA

ABSTRACT

An overview is given of optical communication systems, including fiber, associated components, research trends and economic considerations. Readers are referred to "Optical Fiber Systems: Technology, Design and Applications" by Charles K. Kao, McGraw-Hill (1982) for further discussions of various topics.

TABLE OF CONTENTS

*These lecture notes are edited from the tape and vugraphs of the lectures.
Figures 3.8, 3.9, 3.10, 3.25, 6.5 and 6.10 of this section are reproduced from "Optical Fiber Systems: Technology, Design and Applications" by Charles K. Kao.

1. HISTORY OF OPTICAL COMMUNICATION AND THE ROLE OF OPTICAL FIBER SYSTEMS

1.1 General Background

<u>Why optical</u> "Optical Communication" is as old as smoke signals, semaphore and the use of mirrors to reflect sunlight. In the late 1950's and early 1960's, the invention of the laser, which provides the equivalent of an RF generator at optical frequencies, prompted great interest in the possibility of using optical waves for communication. It is however well to take a broad look at the entire electromagnetic spectrum and ask why optical waves are most suitable for communication. In this context, we should emphasize that optical waves, just like DC currents and radio waves, are just one form of electromagnetic waves; on the other hand, we shall see that the electromagnetic nature of light could not be handled in the same way as that of radio frequency waves and optical signals must be handled very differently.

In principle, information carrying capacity of a carrier increases with frequency. However X-rays and waves of even shorter wavelengths are particulate in nature, therefore subject to noise, which increases as frequency f. Moreover X-rays are attenuated in matter and can not be sent over long distances. The upper limit of useful frequencies is, therefore, the visible or perhaps the long UV. At the low frequency end, frequencies around 1 MHz are already used for radio while VHF and UHF bands at around 1-10 GHz are used for microwave communication systems. The bandwidth available, however, is relatively limited. The IR region is in principle good for communication, but at room temperatures, blackbody radiation peaks at 10 μm, so unless the entire system is cooled, the noise problem would be extremely serious. This then leaves the visible and near IR (say 10^{14} to

10^{15} Hz) as near ideal carriers for communication.

As mentioned earlier, the laser provides a suitable light source at these frequencies. What remains is to devise ways of putting modulation onto the signal and of transmitting the signal over long distances.

<u>Bandwidth requirements</u> Since the major advantage of an optical communication system is its bandwidth, it is useful to have some idea of the bandwidth needed for various purposes. (Table 1)

	Analog	Hifi	Digital
Voice	4 kHz	20 kHz	64 kb/s*
TV	6 MHz	20 MHz	100 Mb/s*
Data			2.4 kb/s-100 Mb/s
Print		30 MHz	

Table 1: Bandwidth requirement for various purposes
*Assuming 8-bit pulse code modulation (PCM)

You will see that a TV picture takes 10^3 times more bandwidth than voice - In other words we confirm here that "A picture is worth a thousand words". However, the market for picture-phones has not yet materialized, perhaps because of our inability yet to make use of such vast amounts of information. At present, data transmission usually works at very low bit rates, 2.4 kb/s to 9.6 kb/s, but if computers are to be linked for interactive data processing, very much higher rates are necessary. There are now experimental data links working at up to 100 Mb/s.

In short, as we move into the Information Age, the demand for bandwidth increases very rapidly, so it is important to have a transmission system which can handle very high bit rates - optical fibers are one such versatile and powerful transmission system, with eventually nearly zero cost and infinite bandwidth. This is not to say that we should discontinue research on other transmission media.

It should also be noticed that different ways of sending the information require different bandwidths. (Table 2). The larger the bandwidth used, the more immune the system is to noise.

Type	Relative Bandwidth
AM	1
FM	3
PAM	2
PWM	6
PCM	16

Table 2: Necessary bandwidth relative to amplitude modulation

1.2 Transmission medium

1.2.1 Free space transmission

This would be simple and probably feasible in outer space, where the signal can propagate without hindrance. For terrestrial propagation, one needs to worry about scattering by dust particles and by density inhomogeneities. First of all, density varies with height in the atmosphere; more importantly, solar heating can cause rapidly varying density fluctuations on a small scale. For example, under extreme conditions, the spot of a laser beam propagated for 1 km in the atmosphere may jitter by up to 1 m. Weather formations can cause serious absorption, e.g. fog can easily cause 40-60 db loss. There is also absorption by atmospheric gases. For example, oxygen absorbs microwaves at tens of GHz. At IR wavelengths, water has strong absorption bands.

In short, free space transmission is on the whole impractical except for a few specialized applications.

1.2.2 Guided propagation

Confocal lens One early attempt (e.g. by Bell Systems) at guided propagation was to use a series of converging lenses, each of focal length f (several to tens of meters) and placed 4f apart. A source 2f in front of one lens is imaged to a point 2f behind it, which is exactly 2f in front of the next lens, and so forth. We studied this system in the 1960's, but found that the propagation was so affected by the atmosphere that even for a distance as small as a few hundred meters, effective propagation was possible only at night, when the air is still.

One can also use a pipe containing a gas whose density is modulated in space (e.g. by heating) to form a distributed gas lens, again in confocal arrangement. The transmission loss can be extremely low and the bandwidth theoretically enormous. In such a system, the gas is on the average denser near the axis and rarer at the edges, so one had in effect created a large-sized graded index fiber.

Metallic waveguides In the 1960's, the H_{01} or TE_{01} mode in circular waveguides, made of metal or helical wires, was intensively studied for microwave frequencies. Such waveguides are possible at optical and IR frequencies; they may in fact be used if we go to the sub-millimeter region. For example, this could be a possible method of transmission at say 10 μm.

Reflecting pipe is another old idea investigated in the 1960's and depends on the high reflectance at near glancing incidence. However, there are many difficulties to make such a transmission line practical.

Dielectric waveguides Fibers were known in the early 1960's, but the loss was high. Two approaches were therefore investigated. One was to use a thin film structure, constructed in such a way that only 10% of the energy was actually in the film, with the remaining 90% in the free space outside the film. This would then reduce losses by a factor of ten. However, it was found that in this configuration even very small bends would cause excessive radiation loss.

The fiber structure is more attractive, provided the material loss could be sufficiently reduced.

The large loss measured in the early works was in fact due to absorption by impurity ions (Fe, Co, Ni, Cr etc) present in the glass; if all these are removed, theoretically it can be shown that material loss can be as low as 1 db/km at 1 μm wavelength. This remaining intrinsic loss is due to Rayleigh scattering by inhomogeneities frozen into the supercooled liquid state.

Because the cross-section for Rayleigh scattering goes as λ^{-4}, there is strong motivation for going to longer wavelengths. In fact there is much current work in trying to find materials and fabrication techniques to achieve fiber structure for transmission at 1.8-4 μm, which would have theoretical losses as low as 0.01 db/km.

1.3 Survey of optical fibers

1.3.1 <u>Material considerations</u> The choice of material must take many factors into consideration: loss, bandwidth, ease of handling and durability.

<u>Loss</u> First of all, one must remove the impurities. This is now no longer a technical problem for SiO_2 based material and in most fibers the only significant impurity is water, but even that can now be reduced to ppb levels. Secondly scattering losses can in principle be reduced by adding different material to control the density inhomogeneity. However, the scattering losses are already so small that there is no great incentive for research in this direction.

Molecular absorption is usually not a problem for pure silica but may be important in cases where dopant is added to change the refractive index. For example boron dopant affects transmission at 1.3 μm and 1.6 μm, so for materials intended for operation at these wavelengths other dopants (e.g. fluorine) should be used. Electronic absorption is normally not a problem since this affects only very short wavelengths.

Bandwidth If the fiber supports a number of different modes, all with different velocities, then the difference in transit time would limit the bandwidth. There are two ways of dealing with this. One method is to use single mode fibers, which is free from this problem and which therefore can have very large bandwidths. Another method is to use graded index fibers which are designed so that various modes have nearly the same velocity.

Secondly, because the light velocity in material changes with frequency, there is also a material contribution to dispersion, which would again limit the bandwidth. However, if the source is truly monochromatic, this would not be a problem. The change of refractive index with wavelength also means that different wavelengths see different refractive index profiles for a graded index fiber and this imposes another restriction on the bandwidth.

Ease of handling For practical purposes, the fiber must be sufficiently strong to withstand normal handling, and must also be easy to join. For example, the jointing problem is difficult for single mode fibers; however my own feeling is that the present problems are more the result of inexperience than of fundamental physical limitations and therefore could be overcome in time. The optical loss in the joints is also an important factor; nearly zero-loss joints are now possible for many types of fibers.

Durability The material must be stable over long times, and must not be subject to fatigue. Glass in an environment containing even very small amounts of water is subject to fatigue, in the sense that its strength deteriorates with time. Obviously this would be a very serious problem for undersea systems, which are on the one hand subject to possible water seepage and on the other to severe stresses when being hauled up for repair.

1.3.2 Basic characteristics

Optical characteristics The values of loss and bandwidth are obviously important, since these ultimately control the complexity of the system that can be supported. The optical properties after concatenation are also important: in principle one should be able to predict the properties of fibers which have been joined together, but in practice this is difficult, especially for fibers made by different manufacturers or on different machines. So at present one can only "play safe" and assume that the effective bandwidth is that for a single length of fiber divided by the number of fibers being joined.

Physical characteristics The small size and weight of fibers, as well as their strength and flexibility, make them attractive for many applications. Moreover, there is the promise that fibers will become very inexpensive, say in the next two years. At that time, there will be much more extensive usage of fibers.

Special characteristics In comparison with conventional communication systems, optical fibers are almost unique in being free from electromagnetic and radio frequency interference. Fibers are moreover not a fire hazard and can withstand relatively high temperatures. These properties have a great impact on the range of applications. In fact it was the possibility of such applications that aroused early interest in optical fibers.

1.4 Possible Applications

The almost infinite bandwidth and potentially low cost of optical fibers render them vital in an "Information Society" in which the primary preoccupation in life is the generation, dissemination and management of information. It is also obvious that optical fiber waveguides, through their vast range of applications, may have a large social impact.

Energy resources Fiber can be important in the energy systems of the future, which will increasingly depend on electricity. In power stations, the electromagnetic power can couple inductively into copper cables, so conventional copper communication systems can be highly hazardous. Fibers will obviously solve this problem.

In addition, fibers can provide a convenient and low-cost

link between power stations, especially if very low-loss repeaterless fibers can be developed to be wired along power lines.

Transportation The movement of people in large cities depends increasingly on mass transit systems in which driverless trains obey computer control. An efficient fiber-based system, by monitoring and communicating to the computer the position of all the trains, will enable the trains to run at both higher speeds and greater frequency, thus lowering transportation costs.

Perhaps more importantly, the ready availability of information through a fiber system may change the travel pattern of the population by eliminating certain journeys. Telephone Conference is already an example of this, as is the familiar Yellow Pages exhortation to "Let your fingers do the walking". Video conferencing is beginning to gain popularity.

Communication This is of course the main area of application. We may envisage improved telephone, data and video services. It may also be possible to provide integrated services to subscribers through a broadband loop plant. Special services may also be available to institutions such as banks, stock brokers and lawyers. Because fiber can provide a transmission medium for different services, it promotes competition, for example between the telephone companies and cable TV suppliers for data and video services.

For conventional communication systems based on copper wires, a very large part of the cost is due to the transmission medium itself. The promise of very low cost optical fibers will eventually have a dramatic effect on network concepts and the mode of operation.

Health service By allowing remote stations to hook into a central hospital well equipped with diagnostic tools, a much greater cross-section of the population can benefit from advanced and expensive equipment. Improved monitoring facilities in intensive care is obviously another application.

Community service In both France and Japan there are now fiber-based systems distributing to the community a large range of information services, including weather, shopping and news. Because of changing family size, increase in leisure time and higher aspirations, much greater use will be made of such information systems.

Military defense One can envisage a large range of both strategic and tactical applications. Fiber provides a broadband information network for C^3I - command, control communication and intelligence. Fiber cable is also unique in being a rapidly deployable; for example a cable laid from a helicopter would immediately establish a communication link. Here, the small size and weight compared to copper cables are crucial properties. It is also possible to use fiber to connect a weapon in flight with a control station in order to provide remote guidance. Lastly, fiber provides secure transmission which is not easily eavedropped upon or tampered with.

Business development Office automation, increased data processing needs and the storage and retrieval of information all demand efficient communication. Fiber can provide a high speed interactive data bus network which is moreover secure and free from interference. Video transmission network based on fibers will permit relatively low cost video conferences.

Entertainment In the U.S., the cable TV operators are providing a large number of transmission channels, something like 100 TV channels in systems coming up in major cities. Since each TV channel occupies 6 MHz, the total amount of information transmitted is enormous, and the need can readily be met by fiber.

In addition to pay TV of the cable type, one can also have a feedback channel for the user to transmit information back to the TV station in an interactive situation.

Education A communication system such as that based on fiber will enable universities to dispense information more efficiently, which is important because not every outstanding scholar is necessarily a good teacher. One can imagine standardized lectures given by experts in teaching being broadcast over a video system, to be used as a core course and supplemented by seminars and discussion groups provided by the teachers catering to the individual needs of each group. This will be important in continuing education and for specialized education, which might demand access to very specialized lecturers. Also a large range of educational material will become widely available.

1.5 Summary

Figure 1 summarizes the main parts of an optical system. The transmitter is an electronics device which uses an electrical signal to drive an optical source. The light is transmitted on a fiber, to reach an optical photo-detector at the receiver, where it is converted back to an electrical signal. The next lecture will focus on the fiber transmission line. Then we shall talk about the other peripheral equipment, and finally about the trend of development.

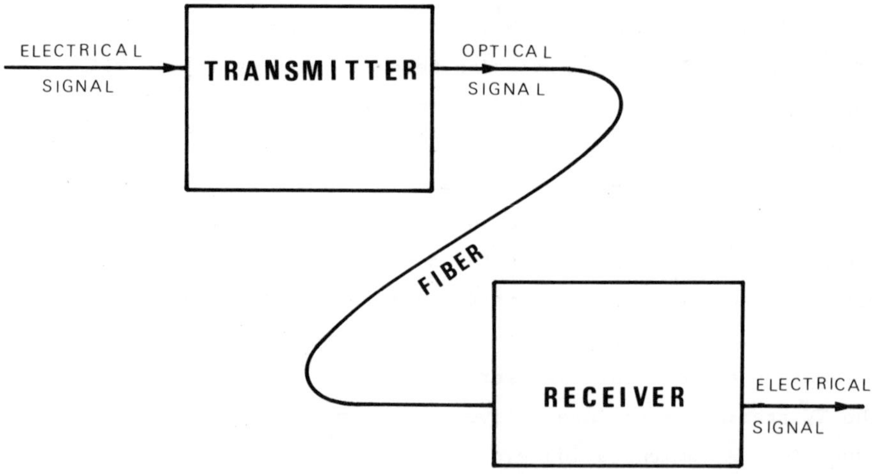

Figure 1: The main parts of an optical communication system.

2. PRINCIPLE AND PRACTICE OF OPTICAL FIBER WAVEGUIDES

2.1 Basic Principles

Electromagnetic theory The basic principles of fiber optics can be understood by solving Maxwell's equations for a cylindrical double-layered structure consisting of core and cladding, and the solution contains all the propagation characteristics. This is a standard boundary value problem, and can be tackled with well known techniques. The work is standard and will not be discussed in detail here, except to quote certain results on waveguide characteristics.

Geometric optics In practice, geometric optics provides a useful and intuitive approximation. Total internal reflection of rays is the basic mechanism which confines light within the fiber. From this simple observation it is already clear that the refractive index of the core (n_1) must be larger than that of the cladding (n_2). Moreover, although most of the energy is confined in the core, there will be evanescent waves in the cladding, which must therefore be sufficiently thick to prevent energy leakage. As we shall see, practical waveguide designs take this into consideration.

Figure 2 shows schematically the propagation of rays within a fiber. All rays with an angle of incidence larger than the critical angle α_c are confined, and are referred to as guided rays which can be related to the waveguide modes. In fact, it can be shown that only certain values of incident angle are allowed, so the waveguide modes are discrete. Rays with incident angle smaller than α_c escape from the fiber as radiative rays, corresponding to a continuum of radiative modes.

There are also the skew rays, which make two angles θ and $\emptyset$ with respect to the local tangent plane and it is possible that one of these angles is below α_c while the other is above α_c.

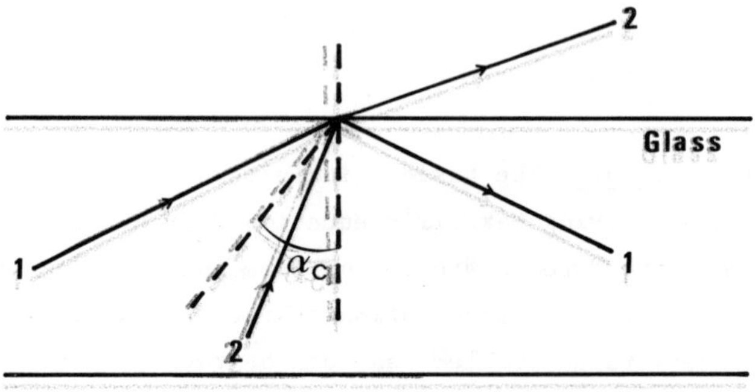

Figure 2: Guided rays (1) and radiative rays (2) in a fiber.

This being the case, such modes are in general "leaky", in the sense that they are continually radiating energy.

2.2 Fiber types

<u>Single mode fiber</u> A single mode fiber is one designed to support only one propagating mode. Figure 3 shows a cross-section of a typical single mode fiber. These fibers are characterized by a small core, about 4 μm, but sometimes going up to 10 μm. The cladding usually has a refractive index lower than the supporting structure. This is because many single mode fibers are made by using a tube, in which the cladding material is deposited first and then the core material is deposited. It is of course possible to make the index of the cladding equal to that of the support structure, which will work equally well.

The relative difference between the refractive index of the core and the cladding is usually discussed in terms of the numerical aperture (N.A.):

$$\text{N.A.} = \sqrt{n_1^2 - n_2^2} \simeq n_1\sqrt{2\Delta} \quad \text{if} \quad n_1 \simeq n_2$$

$$\Delta = \frac{n_1 - n_2}{n_1}$$

For single-mode operation, the core radius a and the numerical aperture must satisfy (for $\Delta \ll 1$)

$$V = n_1 \sqrt{2\Delta}\, k\, a < 2.4$$

k being the radial wave number. This is the reason why the core must be small.

Of course the single-mode fiber really supports two polarization modes. If the fiber cross-section is not exactly circular or if the material is birefringent (which may be stress-induced),

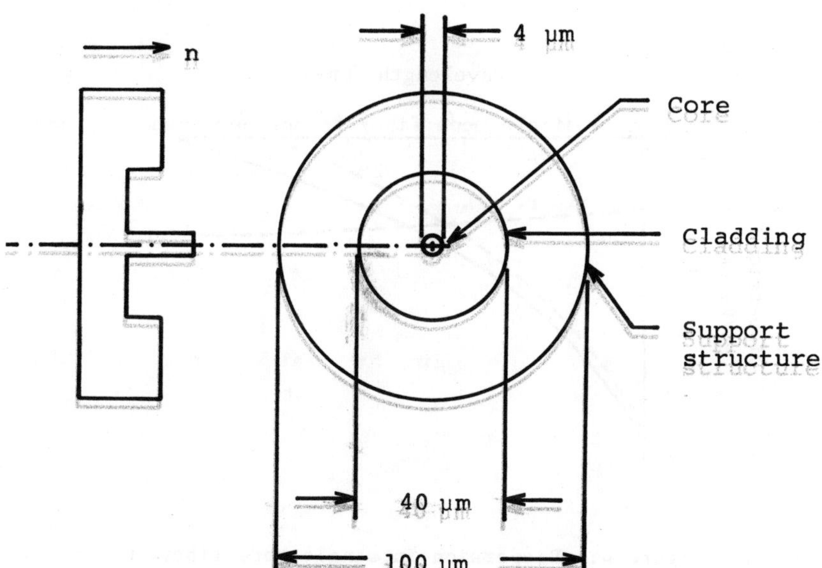

Figure 3: Cross-section of a single-mode fiber.

then the two polarization modes would propagate at different velocities, leading to undesirable signal dispersion. If the fiber is cylindrically symmetric in both geometry and material properties, then the two polarizations propagate at the same speed. But even in this case, because of twists and turns in the fiber, a signal launched with one polarization may be received with a different polarization. This would be a problem if the receiver is polarization sensitive. These effects are important in single-mode fibers since they may be the limiting factors on the usable bandwidth.

The dispersion in single-mode fibers consists of two parts: waveguide dispersion which is always negative and material dispersion which can be positive at long wavelengths (Figure 4). So there is a particular wavelength at which the total dispersion is zero. In order to achieve very high bandwidth, one must choose the material and waveguide geometry so that the zero dispersion point coincides with the operating wavelength.

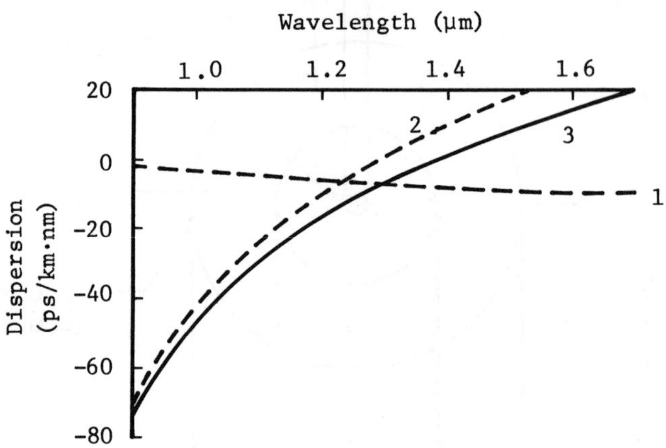

Figure 4: Dispersion in single-mode fiber.
1 = waveguide dispersion,
2 = material dispersion,
3 = total dispersion.

Multimode step-index fiber These are not very popular because they do not have very large bandwidth. Figure 5 shows the cross-section of such a fiber. The core is relatively large, with a diameter of typically 100 μm; the refractive index is uniform within the core. The cladding is relatively thin, so it is necessary to have a fairly large difference in refractive index between the core and cladding in order to avoid radiative loss caused by bending the fiber.

The advantage of such a waveguide lies in its relatively large core and high N.A., which permits optical power to be coupled into it efficiently, thus allowing the use of low cost light sources such as LED's. Moreover, the technical problem

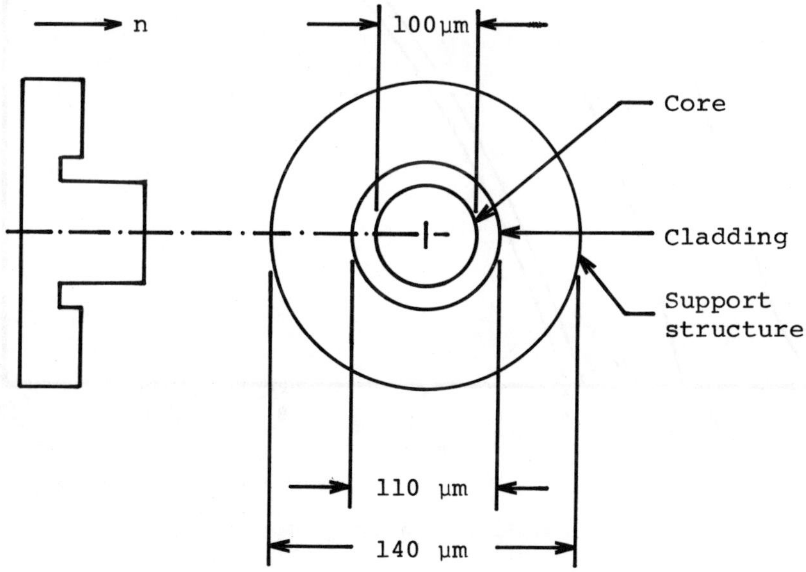

Figure 5: Cross-section of a multimode step-index fiber.

of joints becomes relatively simple on account of the large core. However, because of the large N.A., the loss in these fibers is relatively high, typically 4-6 db/km. Therefore they are particularly suited to short systems. Many computer companies are interested in such a system for computer interface.

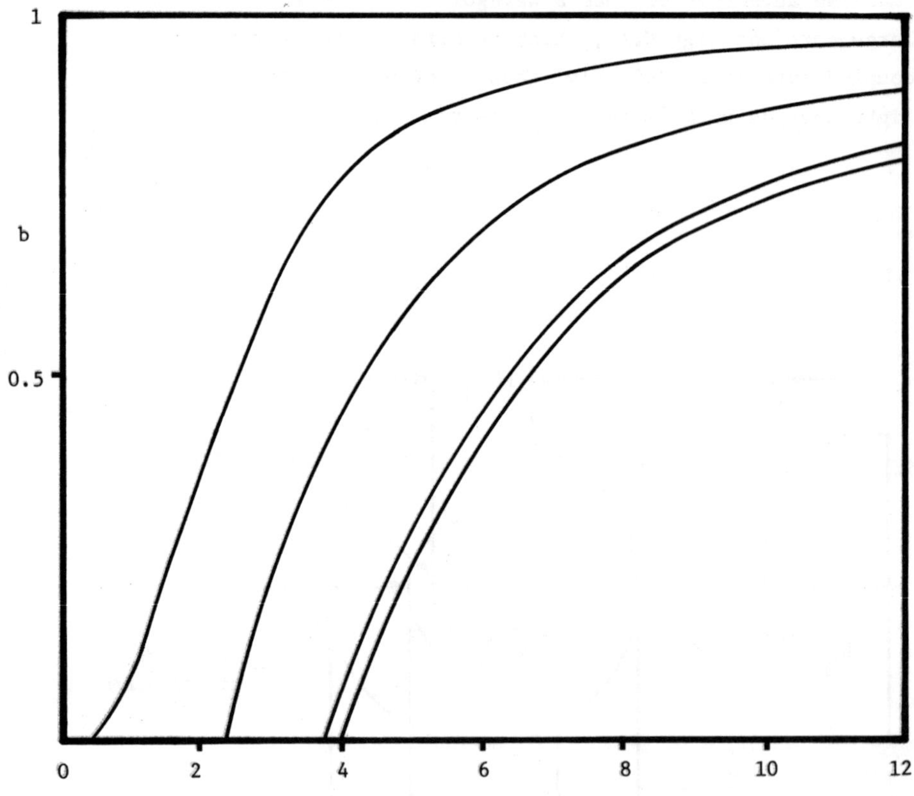

Figure 6: Modes in a step-index fiber.

Figure 6 shows some of the modes in a step-index fiber. In this figure, the horizontal axis is the normalized frequency parameter V

$$V = \frac{2\pi a}{\lambda} \sqrt{n_1^2 - n_2^2} \simeq k\, a\, n_1 \sqrt{2\Delta}$$

Each mode is characterized by a propagation constant β, with mode cut-off given by $\beta = 0$. The vertical axis in Figure 6 is the dimensionless parameter b, related to β by

$$\beta = (b\Delta + 1)\, n_2\, k$$

where k is the wave number.

It is seen that the propagation constant, and hence the velocity, are different for the various modes, so the signal will disperse. This then limits the bandwidth, typically to 20-100 Mb/s over a distance of 1 km.

This figure also shows that if $V < 2.4$, then the fiber supports only a single-mode.

Graded-index fiber The graded-index multimode waveguide is an important transmission medium and probably the best developed fiber (Figure 7). The refractive index of the core assumes a smooth profile, the details of which is crucial in determining the bandwidth capabilities. Theoretically, one can show that the profile can be expressed as a power law, with a power close to 2.

This is now a standard fiber with an internationally agreed set of standard dimensions, e.g. 50 μm core. The numerical aperture varies from country to country, and may be 0.18, 0.2 or 0.23. The cladding must be sufficiently thick, at least a few μm.

The refractive index profile is controlled by adding dopants: germanium, phosphorous etc. In practice, it is common to have a dip in the refractive index at the center of the fiber due to differential evaporation of the dopant material. Such a dip may be detrimental to the performance of the fiber and control of this dip is a key to making consistent fibers.

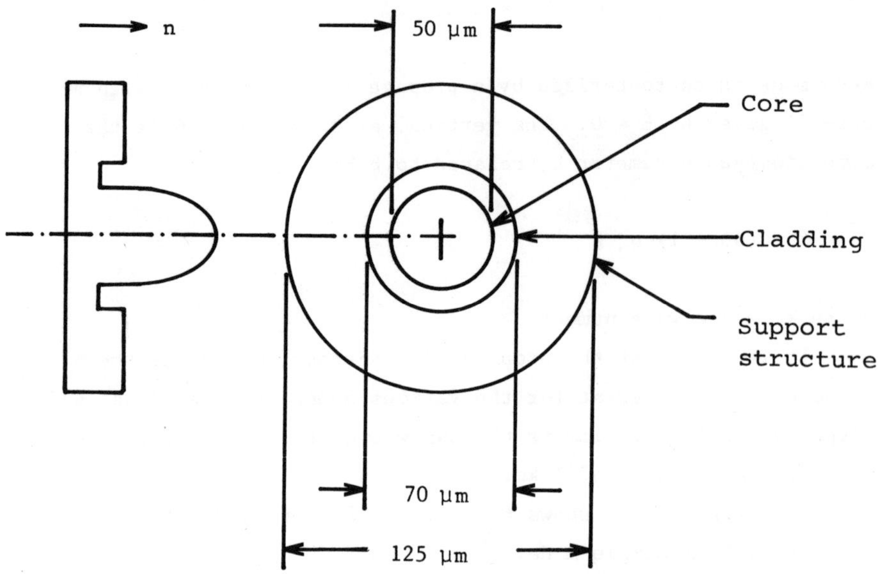

Figure 7: Cross-section of a graded-index multimode fiber.

2.3 Imperfections

In practice, waveguides always have imperfections and it is important to understand the effect they have on signal propagation.

2.3.1 Types of imperfections

Discrete discontinuity For example, a blemish at a particular point in the core will cause scattering and hence mode conversion, both into other guided modes and to radiation modes. In the latter case, there will be a loss in the signal. In the former case there will be an effect on the waveguide dispersion.

Distributed discontinuity These include periodic variation of diameter along the fiber, as well as various microbends. In general distributed discontinuities cause severe losses by radiation, and microbending losses are in practice one of the most serious problems.

The fiber is protected by a plastic coating, which must be designed to minimize microbends in the fiber in the presence of external perturbations. One design uses a soft plastic (e.g. silicone rubber) to surround the fiber, and further protects this with a hard outer layer (e.g. nylon). However, in this case, when the temperature drops, the nylon shrinks by a much larger amount than the glass fiber, thus putting the fiber into compression and perhaps even forcing the fiber to assume a wavy position. This then resembles microbends, and can cause severe radiative losses. The coating structure must be designed to minimize such sensitivity to low temperatures.

Practical discontinuity The fiber diameter may be slightly non-uniform, or there may be a small eccentricity in the cross-section. The material may be slightly birefringent and the refractive index profile may be imperfect. The fibers must also go through certain bends in an actual network. All these imperfections will cause changes in the propagation characteristics.

2.3.2 Effects of imperfections

Increase of loss Imperfections will obviously cause the loss to increase, e.g. by coupling into radiative modes. For single-mode waveguides, the cut-off point for various modes (which is governed by the diameter, refractive index profile and operating wavelength) may be slightly altered by local imperfections. It may happen that a certain segment of the fiber can due to a diameter variation suddenly support two propagating modes, and this would cause very unpredictable loss characteristics.

Bandwidth Usually the increase in loss in multimode waveguides is accompanied by an increase in operating bandwidth. Thus in principle it is possible to artificially induce say mode coupling in order to increase the bandwidth, but this is in general not desirable since imperfections are difficult to control. However, if the refractive index profile is not optimum, or if a single-mode waveguide contains multimode effects (e.g. through a slight eccentricity in the cross-section), then the bandwidth will decrease.

Radiation hardness In general, the optical loss in a fiber increases when the fiber has been irradiated by gamma rays or neutrons; radiation hardness refers to the ability to withstand such radiation without increase in loss. In general, one may expect that if the stochiometry of the material of the fiber is well controlled, then the radiation hardness will be good. Thus compositional non-stochiometry will have an unfavorable effect on radiation hardness, since radiation damages are in the form of electron-hole pairs, which are prevalent at such types of imperfections.

2.4 Fabrication

There are three main techniques for fabricating fibers. Conventional glass making techniques and ion exchange are not commercially exploited on a large scale at present. Most of the high performance fibers are now made with vapor phase reactions.

2.4.1 Multi-element glass fiber formation This is essentially a conventional glass-making technique. The pure core glass in rod form is fed into an inner crucible, while the cladding glass is fed to an outer crucible. The double crucible is heated from the outside to melt the glass, which then flows out through a nozzle at the bottom to form a coaxial structure.

It is possible to produce very long and very uniform fibers by this method, and the technique has been well developed by the British Post Office and by Philips in Holland. One limitation is that the glass used in this process must have a melting temperature which is not too high for the platinum crucible to withstand. The problem is that at high temperatures, impurities in the platinum would contaminate the fiber and would produce undesirable impurity absorption losses.

2.4.2 Ion exchange method There are several companies which make fibers by the so-called selfoc process, in which the core glass contains a highly mobile ion which exchanges with the outside glass, thereby achieving a profile which is suitable for wide bandwidth operation.

There is another process, developed by the Catholic University in Washington. It starts with a Vycor glass, which is a boron doped silica glass. This is first drawn into a rod and is then heat treated so that the glass undergoes a phase separation with certain parts of the glass becoming a crystal. One of the phases is then dissolved by leaching, which removes all the contaminants and leaves a pure silica matrix. The rod is then soaked into a hot solution containing the dopants that are to put back into the glass. Finally, through heat treatment a step-index or graded-index preform is achieved.

2.4.3 Chemical vapor deposition Most of the fibers now being installed are made by chemical vapor deposition, which can further be subdivided into the external deposition technique and the internal deposition technique. After the material has been deposited on a substrate, the substrate is either collapsed or sintered to make a preform.

External deposition Figure 8 shows one external deposition method. The glass forming materials are in liquid form and are carried to the flame by oxygen bubbled through the liquid. The material include silicon tetrachloride which supplies the silicon, germanium tetrachloride which supplies germanium, boron trichloride which supplies boron, as well as fluorine and phosphorous compounds as necessary. The material, carried by oxygen, goes into a burner which is supplied with hydrogen, and the oxy-hydrogen flame supplies the heat to form fine particles of glass in the vapor phase, which condense and coagulate into a soot which is deposited on the mandrel. When a sufficient amount has been formed, the soot can be removed and fused into make a glass preform.

This process can be scaled up industrially, and is therefore economically attractive. Another advantage is that it is unlikely to form a dip in the refractive index profile at the center of the fiber.

One potential problem is the water formed by the oxy-hydrogen flame. However, water can be removed by a chlorine drying process. This is now so well developed that virtually all the water can be removed.

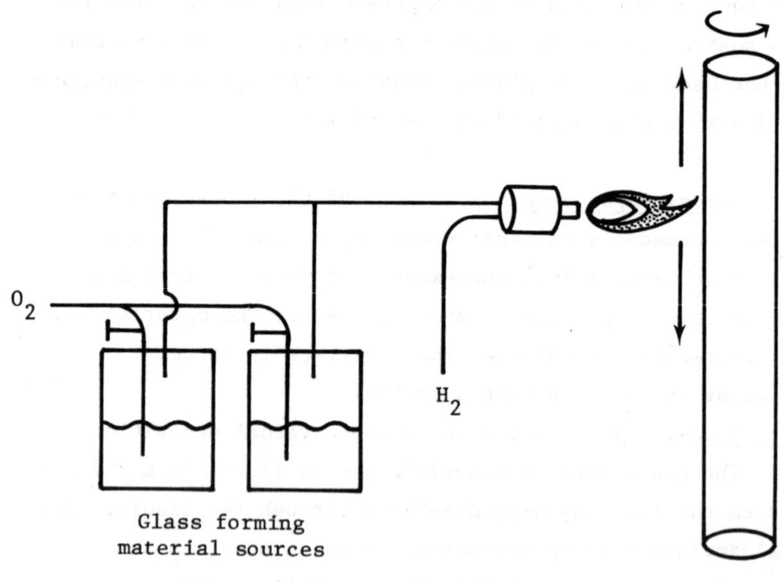

Figure 8: External chemical vapor deposition

A variant is the end-deposition method, in which the flame is applied to the end of the rod, on which layers upon layers of glass are deposited. If the flame is suitably angled and the temperature of the flame suitably adjusted, one can automatically form a graded index profile. This is the so-called vapor axial deposition developed by Japanese scientists.

Internal deposition The substrate is in the form of a tube and the reacting gases are fed into this tube. External heating at one point causes reaction to take place, and the material thus formed will be deposited downstream on the inside surface of the tube.

This process is used throughout the world to make both single-mode and multi-mode fibers and large investments have been made into this process. To ensure good quality of the fiber, the geometry of the tube must be controlled. This in turn allows the temperature of reaction to be constant and the material being deposited controlled. Graded-index fibers with bandwidths of 1 GHz and losses of 3 db/km at 0.85 μm and 1 db/km at 1.3 μm are successfully produced with this method in a number of places.

It is possible to use an internal heating source in the form of a plasma. Microwaves in a cavity create a plasma in a low pressure system. The advantage is that the tube is not heated by the external flame and can therefore maintain its geometry much better. The disadvantage is that because of the low pressure, the amount of material that can be put through, and hence the deposition speed, is somewhat smaller.

An R.F. generated plasma can be used in a high pressure system, allowing a higher deposition speed to be achieved. The control, however, is more difficult for precise rate of deposition and at precise composition.

Fiber drawing To draw the fiber, the preform must first be heated. This can be done either by a resistive or an induction furnace. Flame heating has also been demonstrated, though process control is more difficult.

As the fiber is drawn, the concentricity should be monitored. To ensure the absence of weak spots, an on-line proof test, which involves pre-stretching the fiber, should also be undertaken.

Coating The primary coating may be made of silicon rubber, an UV cured epoxy-acrylate, or a hot melt material which is proprietary to Western Electric. The primary coating is usually applied as the fiber is drawn, and usually involves dip coating.

The secondary coating may be either nylon, hytrel or an UV cured epoxy-acrylate. This may also be applied on-line. The secondary coating is often applied by extrusion.

The coating design is mainly for physical protection. Absorption of water is generally unfavourable to the durability of the fiber, so the coating also serves to alter the resistance of the fiber to water. The coating should also be able to spread any localized external transverse force over a long segment of fiber, in order to avoid damage or microbends. For this purpose the coating dimensions are crucial. Longitudinal forces may damage the coating, so it is necessary to avoid excessive forces during proof-testing.

Coating material should also be able to withstand low temperatures. First of all, shrinkage of the plastic, which may cause microbends, must be avoided. Secondly, the coating must not become brittle at low temperatures.

2.5 Practical waveguide design

Single-mode fibers In this case, the important design criterion is to adjust the geometry so as to compensate for material dispersion, in order to achieve a very large bandwidth.

Multi-mode graded index fibers The refractive index profile may be chosen for operation at a single wavelength (e.g. 0.85 μm or 1.3 μm), at two wavelengths (e.g. 0.85 μm and 1.3 μm), or over a broad spectrum. Current work is concentrated on single wavelength and dual wavelength systems, and broadband operation is not being extensively studied.

Core size and numerical aperture These should be chosen to minimize the bending sensitivity and loss, and to maximize the bandwidth. It is important not to have too high a numerical aperture, otherwise the scattering loss will be undesirably high and the bandwidth will be more difficult to control.

Fiber strength There are two factors which govern fiber strength. Cracks on the fiber surface will of course reduce the strength. Even if there are no cracks to start with, a fiber under tension may be gradually attacked by water, thus forming cracks. This slow failure mechanism is called fatigue.

In practice one must proof-test the fiber and also investigate the dependence of strength on length. In general, the greater the length, the more likely that there will be a low strength point. However, it is possible to have 10 km fibers capable of withstanding a 2% strain.

The strength is dependent on the fiber draw condition, and the cleanliness during the drawing process.

To avoid water attack, one can use a hermetic coating, which may be metallic (e.g. aluminum) or non-metallic (e.g. silicon nitride or silicon oxynitride). The development of these fibers is still very much in the experimental phase.

2.6 Cable Types

Filled cables If an excess stress is accidentally applied to one point of a cable in the field, the cable will break physically; however the stress may propagate along the fiber to cause a fiber break some distance away. This would be extremely inconvenient. A filled fiber ensures that fiber breaks are localized and coincident with the physical cable break. Moreover, the filling helps to prevent water penetration into the fiber.

Tight versus loose designs For applications requiring thin cables, fibers should of course be held tightly within the cable. The fibers in load bearing cables should also be placed in a tight configuration. This affords the fibers some protection from distortions and twists caused by forces within the cable. However, in tight designs the fibers are subject to lateral stress, which would induce changes in attenuation and bandwidth. To take account of this, fibers with a relatively large numerical aperture designed to have low sensitivity to bending are required.

In benign environments it is often advantageous to have a loose arrangement of fibers in the cables. This has several advantages. First of all, there are no stress induced changes in attenuation and bandwidth, so the numerical aperture can be as low as 0.16 if care is taken. Secondly, the low temperature performance is good. Even when properly designed, there may be some thermal shrinkage of the cable over and above the shrinkage of the fiber. The loose design allows the excess length of fiber to be accommodated without causing the fibers to bend through small radii of curvature.

The two main loose designs are the loose tube and the open channel. In the former, the fiber is inserted into a tube slightly bigger than itself; the tubes themselves are however inserted into the cable in a tight geometry with voids filled. The actual cabling is easy, but the fabrication of the fiber/tube unit is more tricky.

In the open channel design, fibers are buffered and placed loosely into slots. The French design uses small channels each holding one fiber, while the Canadian design uses larger channels each holding several fibers. The Canadian design is very flexible. The open channel cables are relatively cheap and quite effective.

2.7 Measurement

Measurement of fiber characteristics is by no means simple. The parameters which should be measured include attenuation, dispersion, bandwidth, refractive index profile and geometry. However, the results often depend on measuring method and conditions, and we briefly discuss some of these problems.

Attenuation measurement typically uses the equilibrium mode launching condition, in which a mode scrambler is necessary. One type of scrambler consists simply of a length of fiber wound on a mandrel; the other type consists of a step-index fiber, followed by a graded-index fiber, followed by yet another step-index fiber. These different scramblers give slightly different attenuation coefficients for the same fiber under test.

When the fiber is wound onto a spool, the tension can cause microbending, which therefore affects the dispersion, bandwidth, as well as attenuation. Some users specify that the fiber must be wound with zero tension, but this creates certain problems of its own.

Profile and geometry measurements are difficult, especially if high precision is required. For example, the Bell System requires that fibers of 125 μm diameter should be measured to ±0.3 μm, which calls for extremely sophisticated techniques.

Direct pulse dispersion measurements and bandwidth measurements by swept frequency technique both depend on the launching condition. Of course, pulse dispersion and bandwidth can be converted into each other, and again the results do not always agree.

In pulse dispersion measurements, there is also the problem of concatenation, namely that results, expressed in 1 km fiber length, from measurements at two different lengths (e.g. 1 km and 2 km) may not correlate with each other.

Geometrical parameters such as cross-sectional dimensions, the numerical aperture, core size and the refractive index profile are usually measured off-line. On the other hand, measurement of diameter, coating concentricity and coating diameter, as well as detection of lumps are performed on-line.

3. ASSOCIATED COMPONENTS FOR OPTICAL FIBER SYSTEMS

3.1 Light Source

We shall briefly discuss the basic characteristics that a light source must have for use in an optical fiber system.

Power output A high power light source can deliver more energy into a fiber, and with fibers of a given attenuation and detectors of a given sensitivity, more power means a greater span length, which is obviously desirable. On the other hand the power carrying capacity of the fiber is limited, so too strong a source would not be appropriate.

Efficiency Low efficiency in converting from electrical to optical power would not only be wasteful of electrical energy, but more importantly, low efficiency would aggravate the problem of heat dissipation and require the source to be cooled.

Life In practice, the expected life of the source is a crucial parameter. Very large numbers of light sources would be required in an optical system. Should sources fail frequently, the cost of replacement (especially labor cost) would be high. In particular, in cases where fiber losses are not low enough to permit repeaterless operation between stations, there would have to be light sources in the repeaters in the field, for which repair or replacement costs would be truly significant. Therefore one typically aims for a minimum of 10^6 hours mean time before failure.

Spectral emission In the first place, the spectral line of the source must match the low loss windows of the fiber. For very wide bandwidth applications using single-mode fibers, the source must be as nearly monochromatic as possible. On the other hand, for graded index fibers, too narrow a line width implies strong coherence among the different modes, giving rise to interference effects commonly referred to as laser noise or modal noise. In this case, the spectral emission should be relatively broad.

It is necessary to put the coherence of the optical carrier wave into proper perspective. When optical communication was first thought of, it might have been thought, by a cursory comparison with radio frequency waves, that the information bandwidth was almost unlimited - if one utilizes only 1% of the operating frequency of 10^{15} Hz, the bandwidth would be 10^{13} Hz. But such a direct analogy with radio frequency would be misleading. The reason is that optical carriers are not sufficiently coherent; thus a realistic modulation technique is intensity modulation. Frequency or phase modulation, depending on absolute coherence, are much more complicated. However, even with this limitation in mind, optical carriers are capable of bandwidths in the many Gb/s region.

A high degree of coherence can only be achieved if the temperature is stable. If the laser is placed on a heat sink and operated at a DC level, then it is not difficult to maintain a constant temperature. However, in the presence of power modulation, especially at low rates, the temperature of the junction would rise and fall with the heat dissipated by the injection current. In this way, the very existence of modulation tends to destroy coherence. However, good coherence may still be possible with external modulators.

In an all optical communication system, in which optical signals are amplified directly, one must also ensure that the amplification medium is tuned to the same wavelength, otherwise the coherence and spectral purity could be lost.

<u>Modulation</u> To put information onto the carrier, one must be able to modulate the source output in a convenient way. Semiconductor lasers can be readily modulated by controlling the injection current.

The rate at which the output can be modulated in a solid state device depends on the inverse of the carrier life time, i.e. the recombination rate. For LED's, modulation frequencies of $\sim$100 MHz can be achieved, while for semiconductor lasers the limit is about 1 GHz. Beyond these frequencies, the dynamical response of the device becomes very complicated. Thus for very high frequencies, one would have to rely on modulation external to the light source.

Operating temperature The source must perform satisfactorily over the temperature range encountered in the actual operating environment. The specifications may be extremely stringent; for example certain military applications require the source to work from $-55^{\circ}C$ to $125^{\circ}C$. This would be quite difficult at the high temperature end. In ordinary communication systems, a working temperature range of $-30^{\circ}C$ to $60^{\circ}C$ may be quite acceptable. An InGaAsP semiconductor laser might not work satisfactorily at $60^{\circ}C$, so a cooling system would be required.

Of course the temperature limits affect the power output, spectral stability and life of the device. This is in fact a major practical consideration, and could give rise to the question of LED's versus lasers for specific applications.

Cost For new applications, the system must be available at a price which the customer can afford. To replace existing services, the cost must be low enough to provide an incentive for changing to a new system. Of course the cost of light sources is only one part of the total cost of a system, and the economics of fiber systems will be discussed in detail later.

Choice of light source It is clear from the above considerations that sources such as tungsten lamps are completely out of the question, on account of their short life, large bulk, difficulties with modulation and so forth. Solid state lasers and gas lasers may be suitable for certain special applications. However, for most purposes, one needs either an LED or a semiconductor laser.

For easy maintenance, LED's are preferred. However the power from an LED that can be coupled to a fiber is limited, so their use is restricted to transmission over short distances. For long distance links requiring few repeaters, semiconductor lasers are preferred.

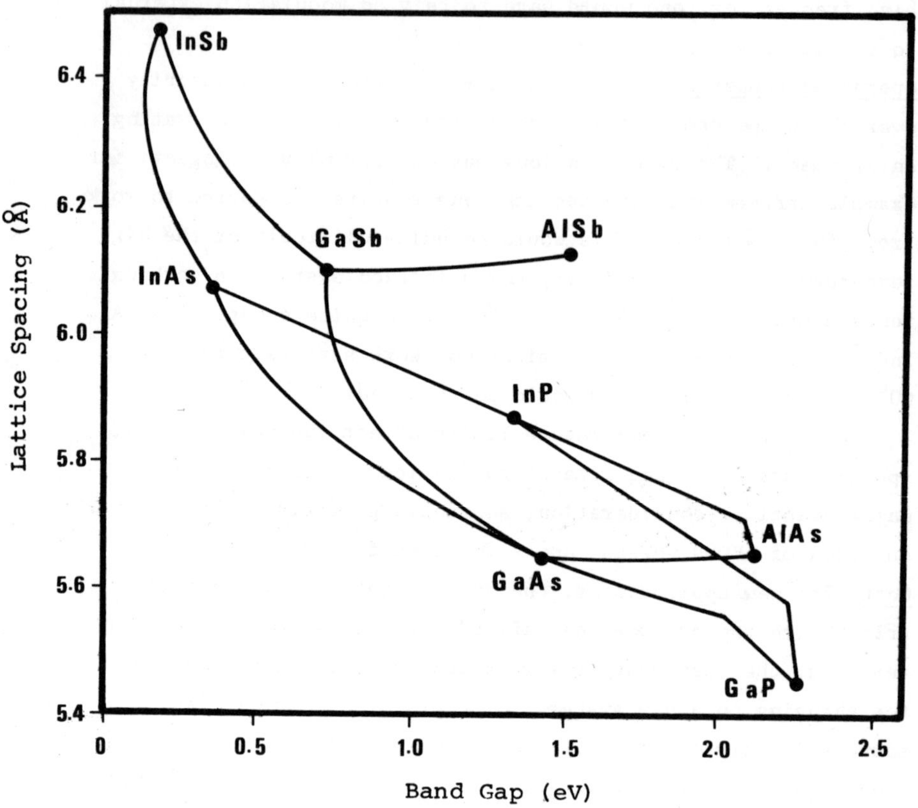

Figure 9: Material chart for III-V compounds.

Choice of material In choosing material for a semiconductor laser, a number of factors must be taken into consideration. In the first place, the bandgap of the material must be such that the optical output falls in one of the low loss windows of a glass fiber, e.g. 0.85 μm, 1.3 μm or 1.55 μm. The lattice spacing should match that of the substrate so that the laser can be made with standard techniques such as liquid phase epitaxy. Finally, the source must be tailored for quantum efficiency, the temperature limits and ease of fabrication. In general III-V compounds are used, although II-VI compounds are also likely to be possible.

It turns out that GaAs and GaAlAs form very efficient and satisfactory systems, and these have been developed in the last decade or so. The fabrication of these devices is now very advanced, and laser life times of 10^6 hours at room temperature are achieved. The temperature range is also good.

Quaternary semiconductor lasers and LED's are also being exploited, e.g. with InGaAsP. Figure 9 is a material chart showing lattice spacing against band gap. It can be seen that by varying the composition, one can choose both a suitable bandgap (say around 1 eV) and also match the lattice spacing to that of the substrate (in this case InP).

3.2 Detectors

The modulated optical wave must be converted back to an electrical signal by a detector. The choice of detector should take into account the quantum efficiency (i.e. the efficiency of liberating photo-electrons), the spectral response (i.e. the sensitivity of the detector to light of different wavelengths) and the frequency response (i.e. the ability of the device to respond to high frequency modulation).

Basic material Three main classes of photovoltaic devices are being used extensively; these are respectively based on germanium, silicon and InGaAsP. The spectral response of germanium (Figure 10) extends from fairly short wavelengths up to about 1.6 μm, so it is possible to make germanium detectors for use at 0.85 μm, 1.3 μm and 1.55 μm.

Silicon response extends just barely over 1 μm (Figure 11) and so silicon cannot be used for 1.3 μm and 1.55 μm detection. This is a pity since the high ionization rate in silicon makes it excellent for avalanche detection.

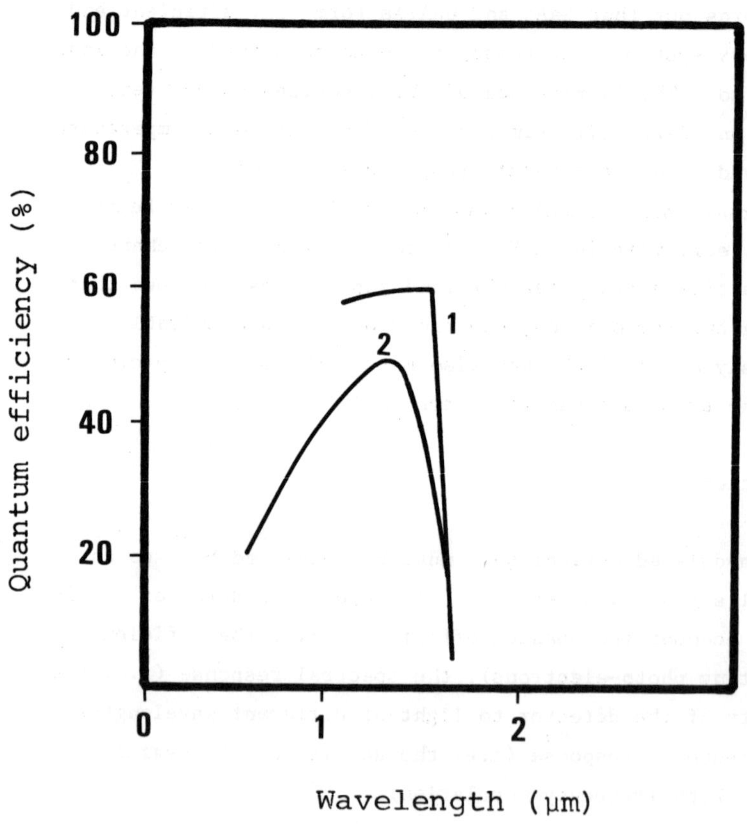

Figure 10: Response of germanium at 300 K
1 = PIN 2 = pn junction

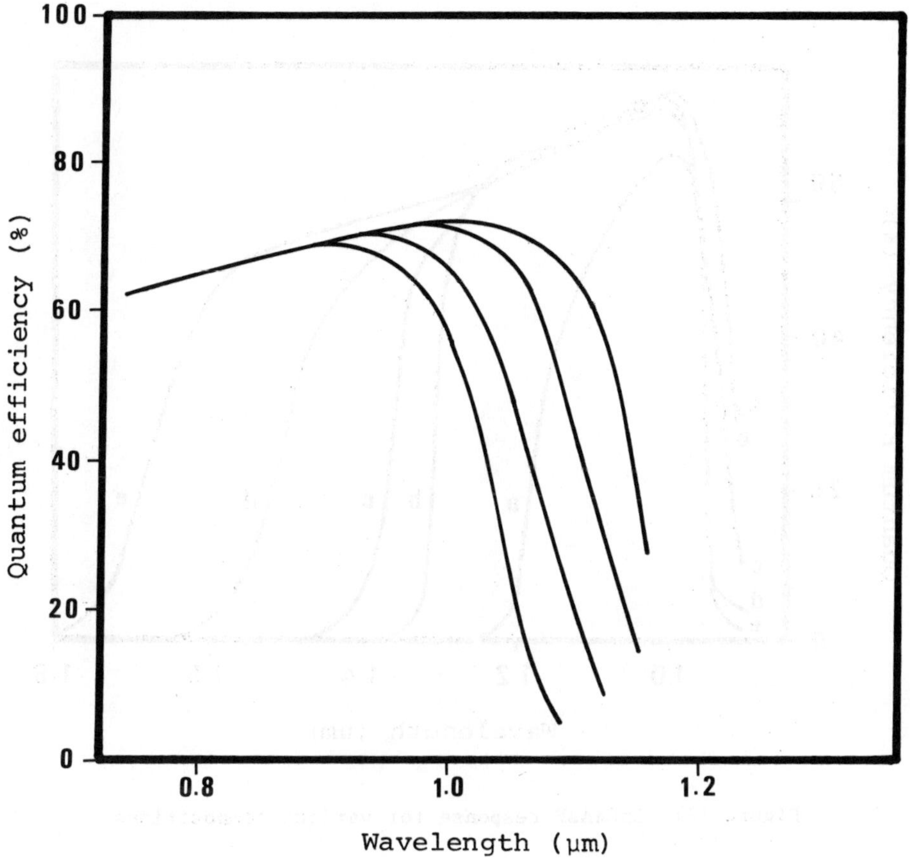

Figure 11: Silicon response for PIN with wide i region of width w. From left to right, w = 0.1, 0.4, 0.8, 3.2 mm.

The response of InGaAsP depends on composition (Figure 12), and response extending up to 1.6 μm or 1.7 μm is possible. It is also possible to tailor the device so that it responds only to one of the wavelengths used (e.g. 1.3 μm but not 1.55 μm). However, there are better ways of obtaining wavelength selection.

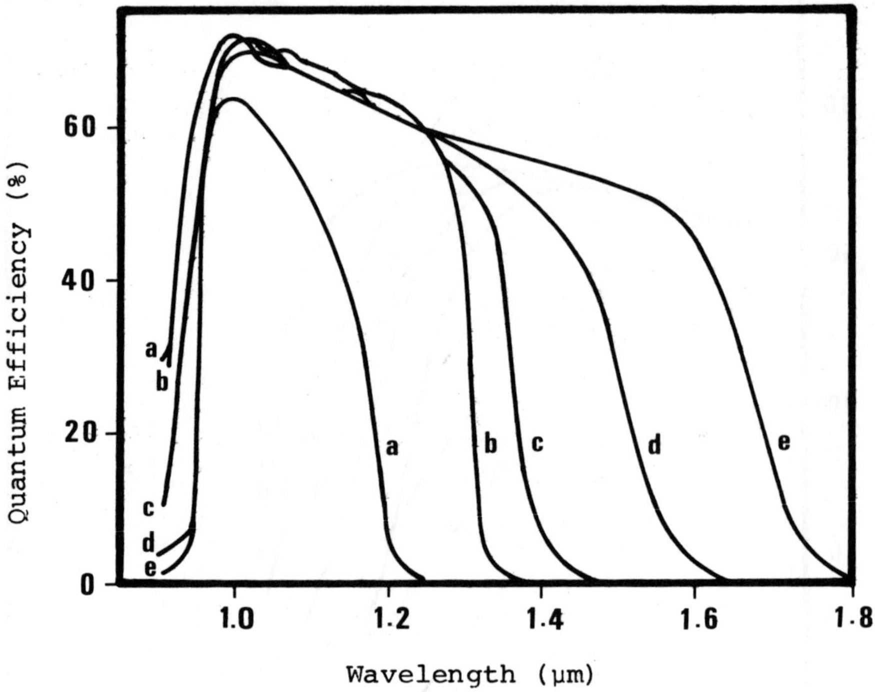

Figure 12: InGaAsP response for various compositions.

<u>Types of detectors</u> There are two main types of detectors: positive-intrinsic-negative (PIN) photodiodes and avalanche photodiodes (APD). In general, if APD's are available, they provide noiseless gain and high sensitivity; unfortunately it is sometimes not possible to make APD's with low multiplication noise.

The excess noise in APD's arise from the randomness of the multiplication process, and thus increases with the ionization probability κ and with the gain G (Figure 13). It is seen that silicon systems ($\kappa = 0.02$) are very much superior to germanium ($\kappa = 1$) or GaAs ($\kappa = 0.1$) for the same gain.

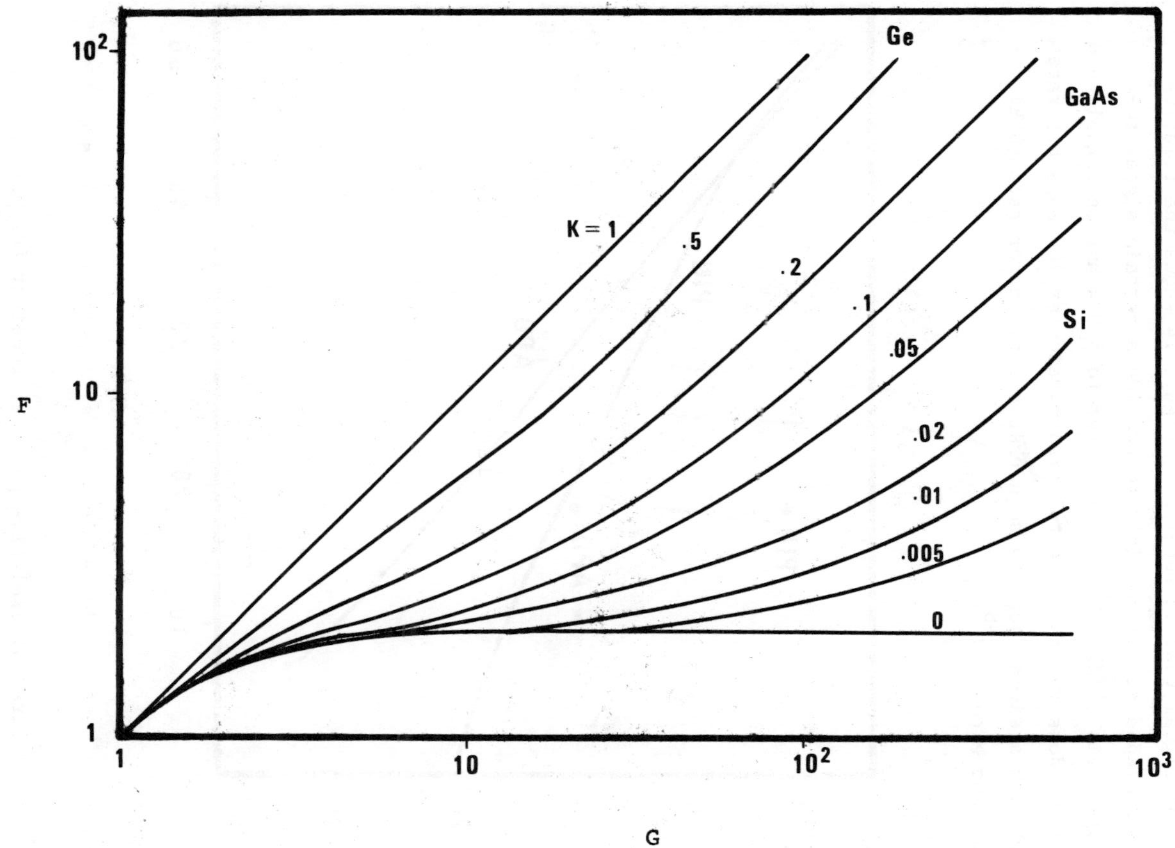

Figure 13: Excess noise factor F vs gain G.

The choice between PIN and APD also depends on the quality of the signal that is required. Figure 14 shows the relative power needed at the receiver to provide a certain signal-to-noise ratio (S/N). The curves refer to 5 MHz analog signals and the two dots refer to digital transmission at 100 Mb/s bit rate and with a signal quality such that the bit error rate (BER) does not exceed 10^{-8}.

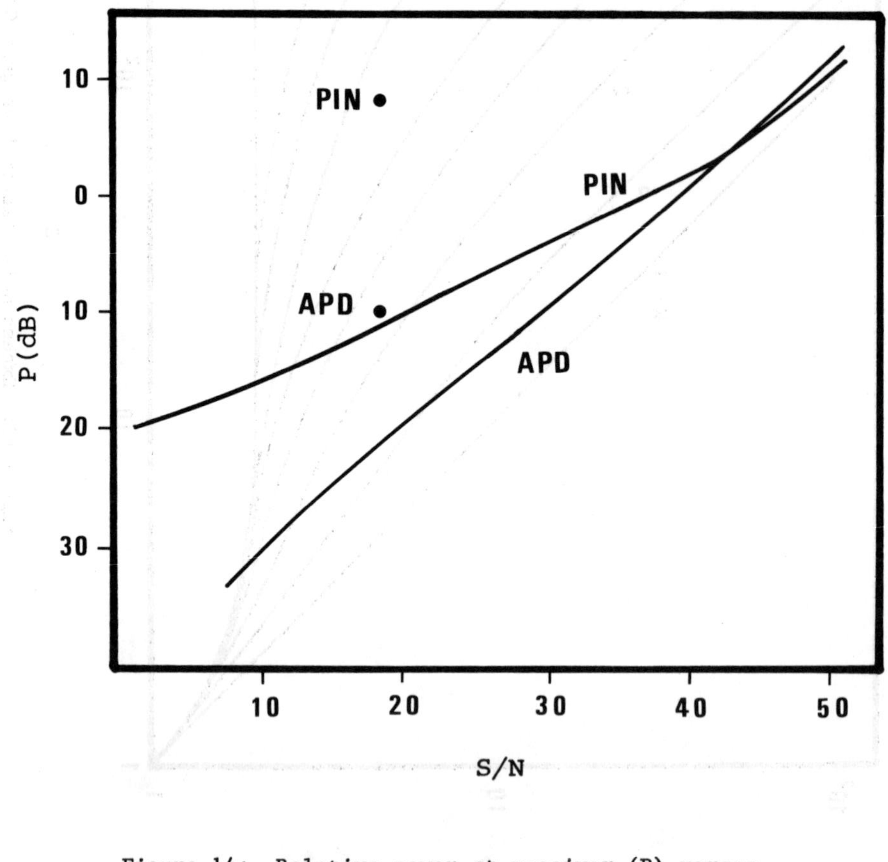

Figure 14: Relative power at receiver (P) versus signal-to-noise ratio for (a) 5 MHz analog signal (curves) and (b) 100 Mb/s, 10^{-8} BER digital signal (dots).

If a high S/N is required, then the received power must be at such a high level that APD and PIN provide the same performance. On the other hand, if the requirement on S/N is not stringent, then APD is much more effective in the sense that a lower received power is adequate.

Amplification The electronics which amplifies the output of the detector will in general introduce thermal noise; thus the design limits are set by the need to reduce thermal noise or achieve noiseless gain. For a pure resistive type of input into the amplifier (Figure 15), the thermal noise is very poor.

One way to improve this is to use the high impedance integrating front end (Figure 16a). This contains an R-C combination (Figure 16b) to integrate the signal. The output from the amplifier goes through an equalizer Z (Figure 16c) which restores the signal by differentiation. This design greatly improves the sensitivity of the amplifier. However, because the front end and the equalizer must match each other, the dynamic range is limited, typically to <20 db.

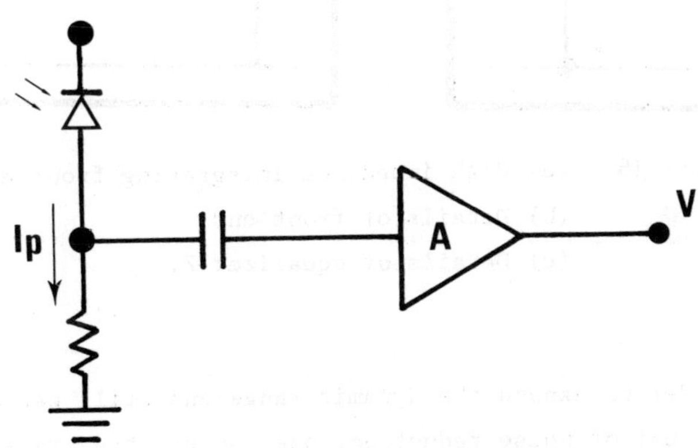

Figure 15: Pure resistive input.

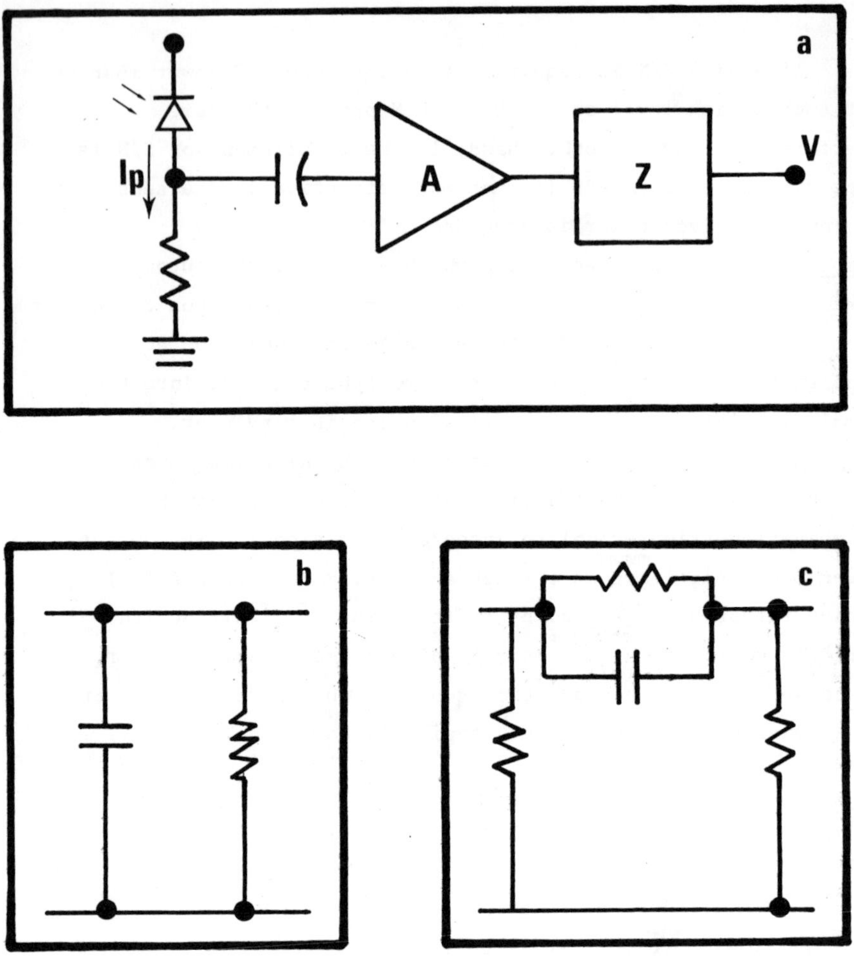

Figure 16: (a) High impedance integrating front end,
(b) Details of front end,
(c) Details of equalizer Z.

In order to expand the dynamic range and still get a certain amount of noise reduction, one can use the trans-impedance circuit (Figure 17), which uses a feedback to improve noise performance.

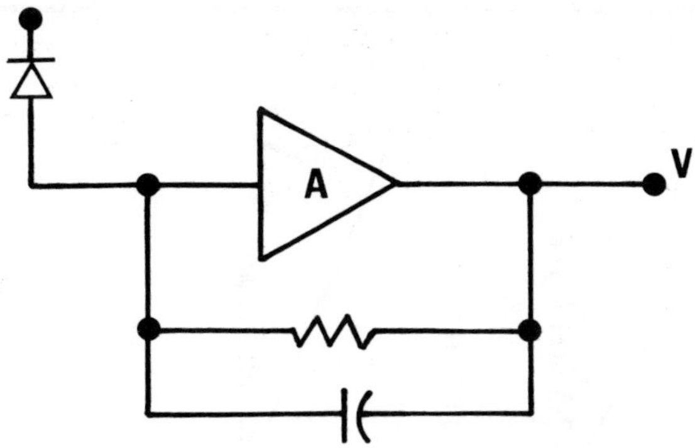

Figure 17: Transimpedance circuit.

Figure 18 summarizes the performance of various detector-amplifier combinations. The graph shows the optical power that must be received by the detector to achieve S/N = 1, for various bandwidths. The curve for an APD followed by a transimpedance amplifier lies the lowest, showing that it is the best combination. For example, for 100 MHz bandwidth, the optical power required is -64 dBm; thus if -44 dBm is actually received, then one can achieve 20 dB S/N. It should also be noticed that the PIN/FET combination is better PIN/BIPOLAR for small bandwidths, but for operation above 50 MHz, the PIN/BIPOLAR combination would definitely be preferred.

3.3 Fiber connectors

Connectors are needed to couple light from the source to the fiber and from the fiber to the detector. Connections are also needed between the fibers; these can be either permanent splices (to be discussed in the next section) or demountable connectors.

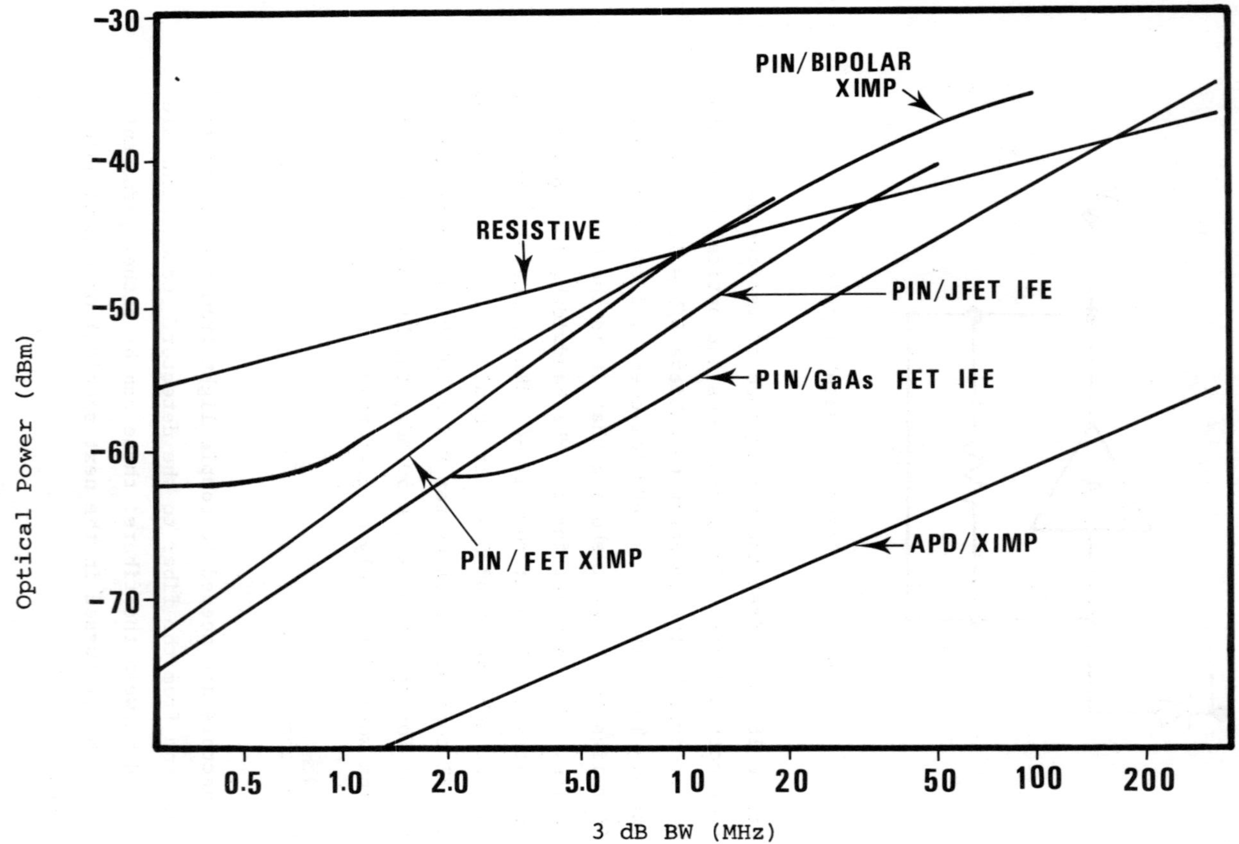

Figure 18: Received power needed for S/N = 1 vs bandwidth.

In a butt joint, the fibers to be connected are simply butted against each other directly. Obviously the ends must be well finished and great care is needed in centering the fibers and ensuring zero separation between the ends. For example, for a step-index fiber, a 20% transverse misalignment (i.e. axis-to-axis separation being 20% of the fiber diameter) will cause an offset loss of approximately 1 dB. For a fiber with N.A. = 0.2, an end separation equal to 20% of the fiber diameter will cause approximately 0.5 dB loss.

In a lens connector, the light coming out from the end of one fiber is focussed by a lens into a parallel beam. The beam is refocussed by a similar lens near the end of the receiving fiber. The relative position between the lens and the fiber is crucial. The lateral alignment is less critical, but the tradeoff is that the angular alignment must be good, e.g. alignment must be better than 1° if the joint loss is to be below 0.2 dB.

The actual connector may utilize molded parts, machined parts or an adjustable mechanical structure. In general, molded parts would be cheapest in mass production while adjustable parts are the most expensive. The ease of assembly, handling and cleaning must also be considered. In systems installed in dirty environments, the joints must be easily disassembled and cleaned.

3.4 Splices

Field splices Two fibers may be joined in the field and the resultant field splice is usually stowed away in a separate compartment. Such splices are generally easy. First of all, there is considerable tolerance on the final appearance of the splice; secondly the splice will not take up any stress after it is stowed away, so there are no stringent requirements on mechanical strength.

Temporary field splices can be made by inserting the two fibers into a V-shaped support and butting them together with the V as a guide. The joint is then held in place by epoxy. Permanent splices must be made by fusion. For ribbon type of cables, multiple splices may be required.

Factory splices Splices are also required in the factory because it is sometimes necessary to make a cable which is longer than the length of a fiber. Factory splices should be totally invisible, so that a continuous coating can be applied. The strength of the joint must be no different from the rest of the fiber. One can then incorporate such a fiber into the cable as though it was one contintuous fiber. These requirements make factory splices more difficult.

3.5 Couplers

Those familiar with microwaves will know of many situations where it is necessary to branch out part of the signal by a directional coupler, which is a four-port device. Similar devices are required at optical wavelengths as well. Couplers are sometimes called tees and stars. A tee is a three port device in which the signal is tapped from the main beam. In a star, many fibers enter a junction, and the signal from any one fiber is distributed to all the other fibers.

Evanescent wave coupler These couplers rely on the fact that the energy in a guided wave is not completely confined within a fiber, but contains an evanescent part extending outside the fiber. A side-arm placed near the fiber will therefore pick up part of the signal (Figure 19). Because the evanescent wave decays exponentially with radial distance from the fiber, this type of coupler does not channel significant amounts of power to the side-arm; it may however be well suited for special applications. For example, in a multimode fiber, the evanescent waves corresponding to the higher modes are stronger and extend further from the fiber. Therefore the side-arm will selectively

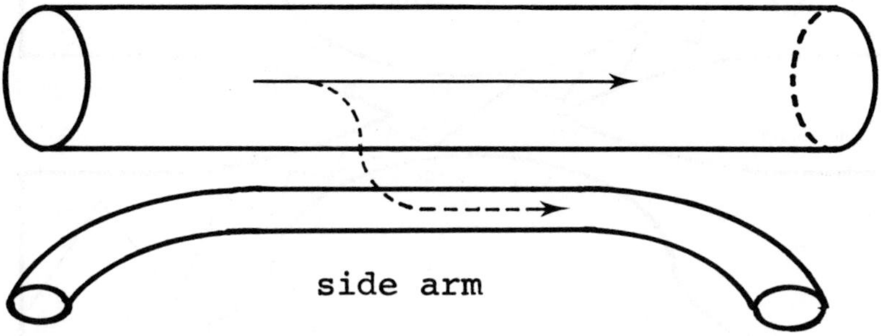

Figure 19: Evanescent wave coupler.

couple out the higher modes. This would be useful in special applications if the different modes are used to carry different signals and one wishes to pick out the signals carried by the higher order modes.

Biconical taper coupler This is a very practical coupler and has been commercially exploited for both the tee and the star configuration. As shown in Figure 20, two fibers are twisted together, the twisted part is heated and then elongated to form a taper, where energy transfer takes place, as illustrated by the ray in Figure 20.

Figure 21 shows the fabrication of a four-input, four-output star coupler. The mounting must be designed so that the coupling region is left free-standing (Figure 22). The packaging material is chosen so that even when the temperature changes, the tapered region is never placed under any strain. Considerable care must be taken on account of differential expansion.

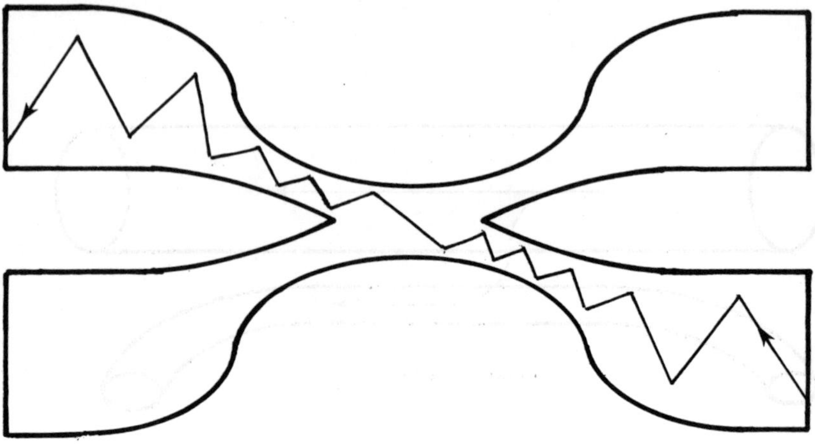

Figure 20: Operation of a bi-taper star.

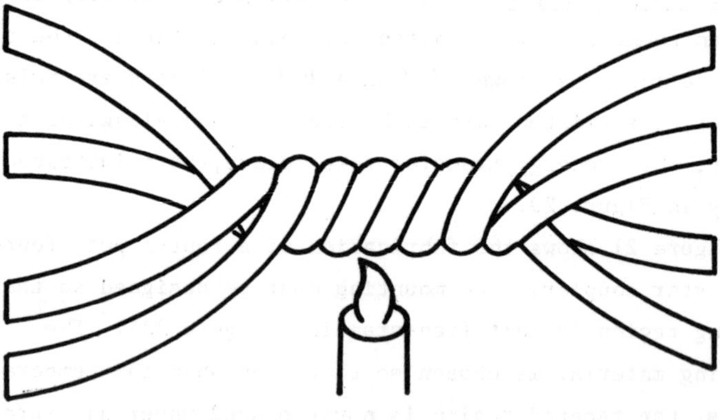

Figure 21: Fabrication of a 4-input, 4-output star coupler.

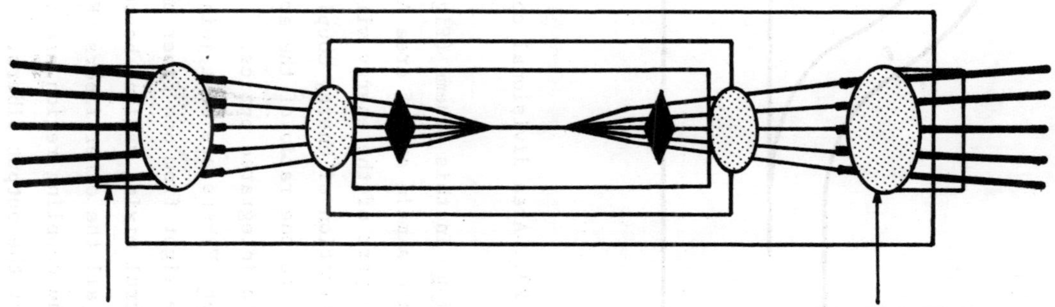

Figure 22: Mounting for 4-input, 4-output star coupler.

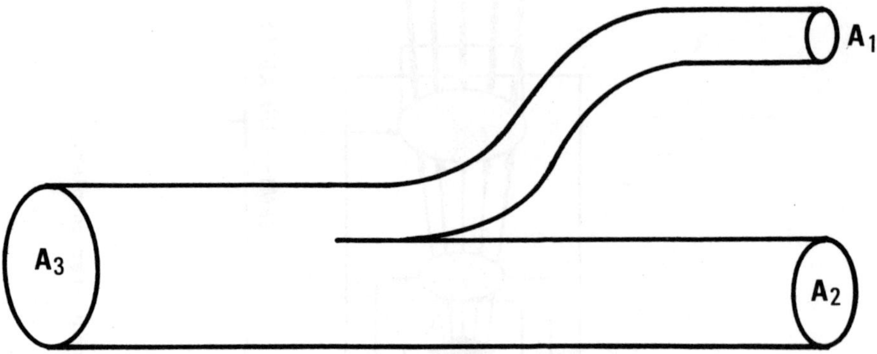

Figure 23: Area divisional coupler.

Area divisional coupling In this scheme (Figure 23), a fiber of area A_3 is split into a main arm of area A_2 and a side-arm of area A_1. Clearly a part of the energy will be channelled to the side-arm; the fraction of energy coupled out is in principle proportional to the ratio of the areas. This coupling method may be useful in integrated optics.

Reflection star coupler Fibers are butt jointed onto a block of large-size fiber. The light from one fiber undergoes multiple reflection in the central mixing region, becoming diffuse and therefore coupling to all the other fibers (Figure 24). The length and shape of the coupling region are important to ensure uniform illumination of the output fibers.

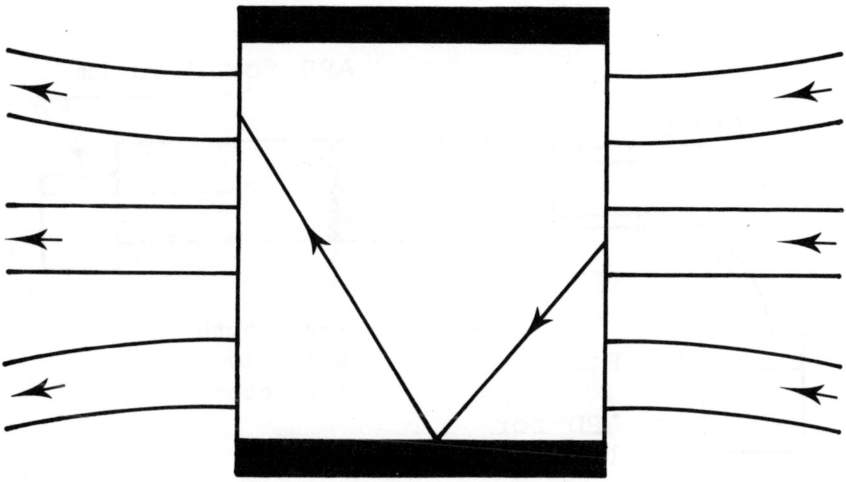

Figure 24: Reflection star coupler.

Dichroic demultiplexing coupler This very ingenious device was first proposed by Japanese scientists (Figure 25), and allows energy carried at two different wavelengths to be channelled to different outputs. The main component is a pair of graded index Selfoc lenses which have the property that an off-axis object is focussed to an off-axis image. The two Selfoc lenses are separated by a wavelength-selective reflector in such a way that one wavelength (say 0.83 μm) is transmitted and another wavelength (say 1.06 μm) is reflected. In the device shown in Figure 25, the 0.83 μm signal goes directly into a detector while the 1.06 μm signal is taken out by a fiber before being coupled to a detector. This device is particularly compact.

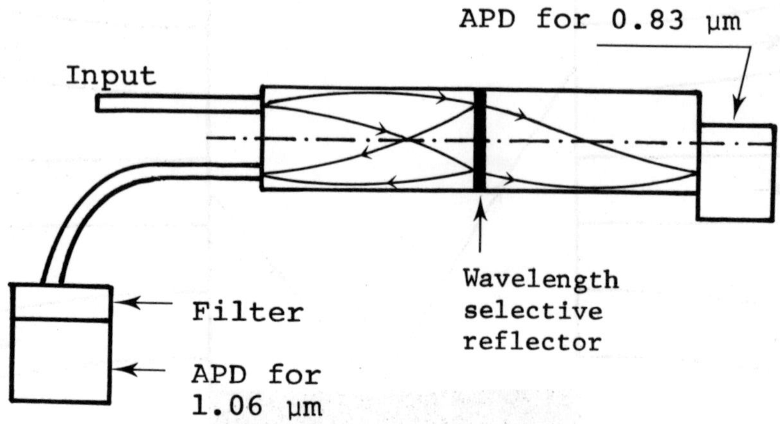

Figure 25: Dichroic demultiplexing coupler.

<u>Other wavelength selective couplers</u> The design concept of a dichroic reflector is simple. Light emanating from the end of a fiber is formed into a parallel beam by a concave reflector. This parallel beam falls on a wavelength selective reflector, so that one wavelength is reflected and the other is transmitted, forming two separate parallel beams. Each of these is refocussed by a concave reflector and falls on the ends of the respective receiving fibers. Unfortunately this device is cumbersome and has low resolution.

The prism coupler works on the same principle as a prism spectrometer. Light from the end of one fiber is passed through a converging lens and forms a parallel beam incident on a prism. Different wavelengths are deviated through different angles and focussed onto the ends of different receiving fibers by an image lens. Again, this system is cumbersome.

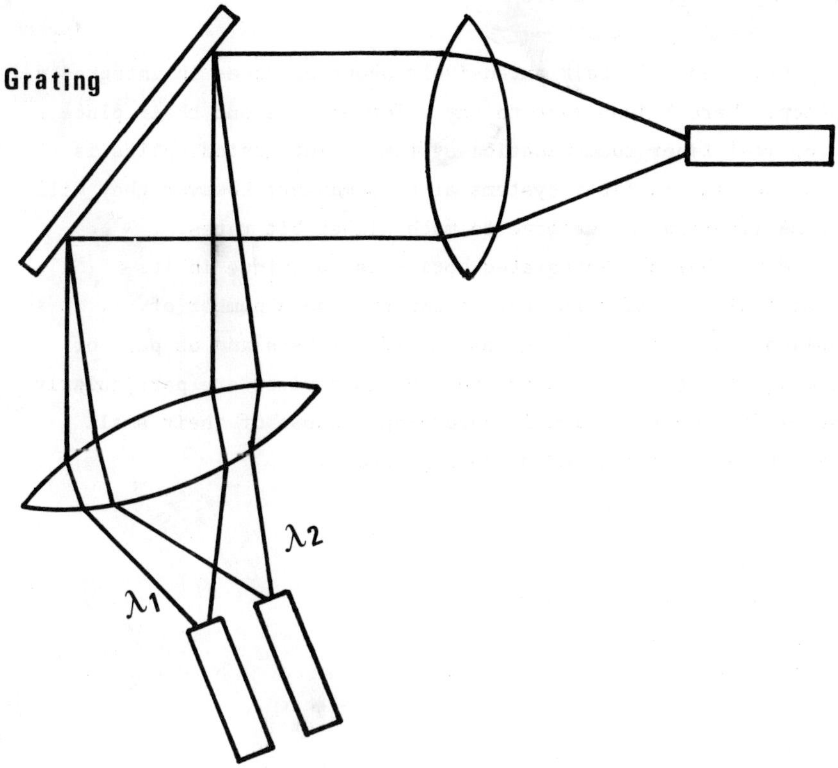

Figure 26: Reflecting grating coupler.

The reflecting grating coupler works on the same principle as a grating spectrometer (Figure 26) and differs from the prism coupler only in using a grating in place of a prism. By using a curved grating, the lenses may be eliminated, resulting in a compact device with high wavelength resolution.

Naturally, all these couplers can be used in a reversible way to multiplex or demultiplex different wavelengths.

3.6 Why Integrated Optics

Dr. Tien will talk extensively about advances in integrated optics. Here I just want to say a few words about their place in optical fiber communication systems. Integrated optics is not essential in fiber systems at this moment; however they will become important as we move to much higher bit rates.

Nevertheless, integrated optics is so unique in its functional capabilities that it can provide a number of applications even now, e.g. as interferometers and as part of receiver front-ends. Integrated optics devices are particularly useful for coherent signal processing because of their small size, high stability and ultra-high speed.

4. RESEARCH TRENDS IN OPTICAL FIBER WAVEGUIDES

4.1 Physical Consideration

Fibers are being used in very demanding physical environments. They are subject to different forces, such as hydrostatic pressure in undersea applications. They should be compact. These requirements impact several areas of research. For this discussion, it is necessary to understand, in the first place, physical characteristics of the fiber, and how changes in these characteristics affect optical performance.

Size The current fiber sizes are already so small that the limit of physical handling is being approached, so there is unlikely to be much greater reduction. The size however influences the optical performance: a relatively large cladding-to-core ratio is needed to keep bend losses to an acceptable value. Of course, cladding-to-core ratio can be traded off against numerical aperture. For example, a 50 μm core, 125 μm cladding fiber will perform satisfactorily with N.A.=0.2, but for 100 μm core, 140 μm cladding, N.A.=0.3 would be required.

A larger N.A. helps to keep the energy within the fiber, and bend losses are generally acceptable for N.A.$\geq$0.2, and become almost negligible for N.A.$\geq$0.25. However, to achieve a large N.A. requires addition of dopants, thus increasing losses. The dopant contribution to loss is usually negligible at N.A.=0.2, but increases to 0.2-0.3 dB/km at N.A.=0.25. For large N.A., the necessary refractive index profile is also difficult to control, thus limiting the bandwidth. Therefore at this time, it appears to be a good compromise to have N.A. about 0.2, and a fairly large cladding-to-core ratio.

Flexibility Fibers must go through various physical treatment and so must withstand bending and stretching. There are also issues related to component design. For example, a length of fiber may be clamped inside a connector, but this may cause problems with differential expansion. One solution is to deliberately design a bend in the portion of the fiber inside the connector; however this induces a permanent stress. All these require better understanding of the physical characteristics of fibers.

Volume Current world production of fiber is probably under 10^6 km/yr, but will be dramatically scaled up in the near future, perhaps by a factor of 10^3. One must then pay attention to resource considerations. For example, germanium is used as a principal dopant. In the early days of the semiconductor industry, $GeCl_4$ was in great demand and hence widely manufactured. But with silicon taking over from germanium, $GeCl_4$ supplies have declined. Thus, for a time, it was difficult to obtain $GeCl_4$ for doping; this situation has improved because the industry has been stimulated by the demands of fiber manufacturing. Another dopant is fluorine, but when used in very large volumes, its effect as an environmental pollutant must be taken into consideration.

Structure There is a debate on the relative merits of single mode fibers versus multimode fibers. Research and development will no doubt lead to improvements in the performance of both types and it is appropriate to expect that each will find its own area of application and both will coexist for some time to come.

There is also research on large cross-section cables containing very many fibers. Currently, each cable contains from 1 to 20 or 30 fibers. Bell Systems has developed a cable with 144 fibers and cables with up to 3000 fibers have been demonstrated. One area of research is to develop a fiber with multiple cores, which would then have the functional capabilities of several fibers but the handling convenience could be irretrievably lost.

4.2 Material Considerations

At present, fibers are based on silica. This is an important system, and there will be continuing material research in this area. For example, if water could be completely removed, then new wavelength windows would open up, particularly in the infrared.

One would also want to understand more fully the effect of dopants such as fluorine, germanium and boron on optical absorption.

However, the molecular absorption band of SiO_2 limits its use to wavelengths below 2 μm and therefore prevents the exploitation of low Rayleigh scattering cross-sections at long wavelengths to make very low loss fibers. There is therefore some research on making fluoride glasses, whose molecular absorption bands would be in the 3-10 μm IR region. It is hoped that transmission windows up to 5 μm can be opened, with losses as low as 0.01 dB/km. Such low loss would permit repeaterless links over thousands of km. Such long distance systems are big enough and expensive enough to warrant work on materials for long wavelength operation.

Many material scientists are now engaged in research on new materials for making fibers, and several classes of materials have been proposed as good candidates. There are, however, a number of difficulties. To control the material, one often wishes to add a third component to a binary glass-forming compound; however, the molecular absorption band of the resultant material is difficult to predict in advance. Furthermore, particular compositions may or may not form a glass. Even if the glass can be formed, detailed experiments, demanding much time and money, is required to check whether the material is durable, whether it is prone to spontaneous recrystallization and whether it is water soluble. For example, KCl crystals have good IR transparency, but are too hygroscopic to be practical. One must also consider the possible toxicity of the glass, not only by itself, but also in the event of accidental chemical reactions. The material must also be easy to purify; this excludes any substance with a high affinity for ions which would introduce absorption bands at the wavelength of interest. The abundance of the material is also a factor to be considered.

The ease of fabrication raises several further issues. Can the material be conveniently purified on a large scale? What types of reaction can be used to form the material? For example, the vapor phase reaction for making SiO_2 may not be appropriate for other materials. Finally, to achieve a coaxial structure, the core material must be compatible with the cladding material.

All these considerations put stringent limits on the choice of material for infrared fibers.

4.3 Application Considerations

The direction of research must of course be guided by the possibilities of future applications.

Repeater spacing Systems with long repeater spacing would of course be desirable and this is the impetus for research on low loss infrared fibers.

Gain possibilities Loss in the fibers must be countered by gain in the system. At present the gain is provided by repeaters which detect the optical signal, converts it to an electrical signal amplifies it and then reconverts to an optical output. Direct optical amplification is possible, but unlikely for several reasons. First of all, optical amplifiers are wide band devices; the signal, occupying perhaps only 1 GHz, is amplified along with a wide spectrum of noise, and signal quality is degraded. Secondly, optical gains saturate very quickly, so large power gains are difficult to achieve. Perhaps the only possibility will be in providing compensational gain. For example, a switch with 3dB loss can be followed by an optical amplifier with 3 dB gain. Over narrow ranges of signal-to-noise ratios, optical amplifiers can work satisfactorily. They will probably be useful in an integrated optics setting, where they can be combined with switches, couplers etc. to provide lossless devices.

Dispersion limit In single mode fibers, one normally wishes to operate at a point of zero dispersion. For a given material, one can tailor the fiber geometry to shift the zero dispersion point to the operating wavelength, but such shifts often imply geometry and numerical aperture values which tend to increase the loss. It is therefore an interesting area of research to investigate whether addition of a third or fourth component to a binary glass compound can change the material dispersion in a desired way without introducing extra loss.

For multimode fibers, it has been suggested that instead of zero dispersion at a single wavelength λ, one can aim for very low dispersion over a range from λ_1 to λ_2. Such a system would be able to support many large bandwidth channels. This is a promising area of research.

Radiation hardness Research on the effects of radiation should be directed not only on the hardness of currently used fibers, but should also address more fundamental issues such as the damage mechanism and whether radiation damage can be associated with non-stoichiometry.

If radiation damage depends on stoichiometry, one would have the interesting possibility of using radiation hardness as a measure of homogeneity achieved in the material, and of evaluating different fabrication processes in this regard. For example, SiO_2 made by flame hydrolysis contains water, and is found to have improved radiation hardness. According to one theory, the formation of SiO_2 from Si and O proceeds through SiOH, and the OH has a refining action on the stoichiometry. This is an exciting area because we know so little about the reaction dynamics and how to control stoichiometry.

Environmental limits Research is also needed to study fibers which can withstand severe environmental conditions such as extreme temperatures. Space and transoceanic application would also require special fibers.

4.4 Fabrication Technology

Crystal growth If one chooses a crystalline material, the technology associated with crystal growth must be studied. Very often, the rate of growth is too slow and one must devise ways of speeding it up. The dendrilitic growth method has been shown to be a very fast (a few mm/sec) way of making long crystal fibers. The method consists of inducing crystallization to start at the edge of a wheel and pulling the seed.

Whatever fabrication procedure is adopted, the resulting crystalline fiber must not have so many dislocations as to cause serious scattering loss, or too many imperfections affecting the strength. Cladding must also be considered. At present, plastic is used to clad crystalline fibers.

Chemical vapor deposition If this technique is to be used, then the source material must be available in a volatile form, either in liquid form at room temperature, or in a solid form which can be vaporized at some convenient temperature. For instance, if aluminum is to be used, a solid source is necessary.

Chemical vapor deposition is attractive because of its fractional distillation feasibility. If the required material is evaporating from the liquid preferentially than other materials, then there is a natural purification process going on. This is the main advantage of this technique, which allows us to make silica fibers of such high purity.

The reaction dynamics of the chemical deposition process have been extensively studied; the results, however, are still fragmentary and we expect new results in this area to be forthcoming which would be more comprehensive.

Glass forming There is renewed interest in glass forming techniques because one wants to make new types of glass, e.g. fluoride glass for long infrared transmission. First of all, one needs to determine, for various systems, the range of compositions which permits glass formation. One also finds that many materials suitable for infrared use cannot be made into a glassy state except through very rapid quenching. This is not too stringent a requirement for fiber making, since the fiber pulled from a crucible nozzle is indeed cooled very rapidly on account of its very small diameter. Finally one needs to study whether the glass is subject to spontaneous or long-term recrystallization.

Coating design The coating may react with the fiber. For example, certain types of polymer coating form links with the fiber surface. These can be used advantageously in controlling water attack.

The development of hermetic coating is a fertile area for further work. There are a number of ways of applying the coating. The material may be evaporated onto the fiber surface; in this case the material settles quite gently onto the cold or slightly warm surface, and little damage is caused. The material may also be applied by sputtering, but the high energy can cause damage to the fiber surface. Dip coating and chemical vapor deposition can also be used. In the latter technique, the chemical is deposited on the fiber and decomposed in situ to form a second phase which is hermetic. For example, if an organic polymer containing metal is deposited upon heating, reduction occurs leaving a metal layer. The materials used for hermetic coating include silicon oxynitride, graphite and metal (principally aluminum).

The development of hermetic coatings is still in its infancy and much remains to be explored.

4.5 Status Of Silica Systems

Quality We now have sufficient control over the optical performance and physical performance of silica fibers for them to be used in most applications. However, for some special applications, silica may not meet the requirements.

Quantity The prospect of very large scale applications necessitates attention to problems associated with quantity production. It is necessary to reduce costs, from the current level of US 30-40¢/m to US 10¢/m or lower, e.g. by exploiting economy of scale and by automation. Automation would not only cut down on labor costs, but would improve quality control by eliminating variation introduced by different human operators.

It is necessary to understand the factors which limit the rate of deposition and the rate of coating. The problem of pollution must also be dealt with as quantity is scaled up.

Fiber tolerance limits cannot be measured down to fraction of a μm by conventional optical techniques. Convenient indirect methods must be devised which are suitable for large scale use.

Bandwidth limits The limits on the bandwidth should be further studied. For example, material dispersion should be better understood and the effect of dopants needs to be clarified. The fiber will exhibit variations (e.g. in diameter) within certain tolerance limits; it is necessary to understand how such variations affect the bandwidth. For example, it has been shown that periodic variations in the fiber diameter, provided they are of short wavelength, induce non-radiative mode conversion and thereby increase the bandwidth.

Polarization maintenance In applications with polarization-sensitive detection, it is necessary to have fibers which maintain the polarization of the field. There are two opposite approaches to this problem. The first method attempts to make the fiber highly uniform, tension-free and isotropic. The drawing condition is carefully controlled so that no birefringence is induced by twists etc. This has two effects. First of all, good isotropy would suppress coupling between the two polarizations. Secondly, because of the near degeneracy between the two polarizations, energy transfer between the two, which is directly analogous to beats, has a low beat frequency, and hence a long beat wavelength. In other words, energy transfer occurs only over very long distances.

In the opposite approach, one deliberately splits the two polarizations, either by using an elliptic core, or by using an elliptic cladding, which on account of differential expansion, induces birefringence in the core. The principle is analogous to the well known property in quantum mechanics: widely separated levels do not mix.

4.6 Fatigue Resistance

Fibers tend to fail when placed under tension and in the presence of water. This is called fatigue and there is some current work on trying to understand the origin of fatigue and devising ways of preventing it.

The phenomenon of fatigue is now fairly well characterized. For example, at a tensile stress of 5 GPa, the life of a fiber is typically only 1 second. For lower values of stress σ, the life t of the fiber is found to improve, roughly as

$$t \propto \sigma^{-24}$$

Thus at low values of stress, the life can be very good.

It is now believed that water attacks the fiber surface, causing weak bonds which are broken by tension. Therefore a hermetic coating is one method of providing resistance against fatigue. Another interesting method is to use a skin layer on the fiber itself. In this way, it is possible to place the skin under compression, by as much as 1%, when the fiber is not under external load. If the external tension is not too large, the skin remains in compression and is therefore not subject to water attack. However, the fiber fabrication process becomes much more difficult, so this method may not be practically attractive.

4.7 Conclusion

Development of new fiber types must pay attention to bandwidth, attenuation, microbend sensitivity, temperature and geometry. These parameters will determine the trade-off between different fiber types. Bandwidth will be particularly important. If one envisages 10^3 km repeaterless links operating at 1 GHz, the necessary bandwidth is extremely high.

Research and development must be influenced by the applications. In this respect, it is important to distinguish between large volume applications such as telephone systems and unique applications such as sensors. The timing is also important, since needs change as a function of time.

5. RESEARCH TRENDS IN ASSOCIATED COMPONENTS

5.1 Connectors and Splices

Connectors Different types of connectors will be needed for different applications. Plastic molded types will be suitable where large numbers of connectors are needed. The ferrule type, with fiber sealed in precision tubes, is needed for high quality interconnects with very low loss. Adjustable connectors are probably demanded for connecting single mode fibers. Connectors designed to be durable, cleanable and rugged are needed for special applications.

Research in connectors really falls into the realm of mechanical engineering, and deals with questions such as physical robustness and cost-effectiveness. Perhaps the only material-related problem is to investigate the possibility of high precision plastic molds. Good results have been achieved with plastic loaded with silica dust.

Splices Research can be divided into several areas: fused splices for factory use, fused splices for field use and quick temporary splices for field use. Splices are now highly developed and very low losses are achievable, even for single mode fibers. Fiber dimensions can now be controlled very well, so that provided alignment is good, there would be little attenuation across the splice. Again current research on splices is mostly mechanical in nature.

5.2 Cables

Small count cables, i.e. cables containing a small number of fibers, are already quite well developed and there is not much research except to devise more cost-effective fabrication methods, e.g. by adopting various well known techniques from the making of conventional cables. Thus, one focus of cable research is on large count cables, i.e. those carrying a large number of fibers.

Large count cables are needed for long distance trunk links, and in distribution networks in which signals must be branched to each subscriber. Here the problem is simply one of splaying N fibers in one cable out to cables with N_1, N_2, ... fibers, until there is a selected number of fibers for each subscriber. Some design work and cable fabrication technology work will be carried out, but again the work involves rather standard engineering.

The prospect of very low fiber costs opens up the interesting possibility of spatial multiplexing. The signal is somehow split up, transmitted along several fibers at lower bit rates and reassembled at the receiving end. Such a scheme would involve large count cables. A number of problems remain to be solved; for example the different delay times along different fibers have to be reconciled.

Signal processing is another interesting application for large count cables. Even in the 1940's, fibers were used to transmit optical images directly. A large number of fibers, each corresponding to a picture element, were assembled into a coherent bundle. The output from the bundle would reproduce, element by element, the picture "seen" at the input end. Unfortunately, high losses limited these early applications to span lengths of not much more than a few meters . The low loss fibers now available open up the possibility of similar applications over much longer distances. Coupled with arrays of light sources and detectors, new information processing methods could be envisaged.

5.3 Couplers

One area of research is the development of wavelength selective couplers for window multiplexing, i.e. to couple into one fiber signals transmitted at different wavelength windows (e.g. 0.85 μm, 1.3 μm, 1.55 μm). There is also the interesting problem of wavelength multiplexing within the same window (e.g. using 0.81 μm, 0.82 μm, ... within the window nominally referred to as

0.85 μm window). The technological problem that must be overcome is the stabilization of the laser wavelength. This problem may put a limit on the minimum spacing between channels of perhaps 0.1 μm, against a theoretical spacing limit of ~2nm. This area is still not well characterised, and much development is expected.

Multiport couplers will be needed for data bus; this will be discussed later.

Bidirectional couplers are needed for systems in which a single fiber is used to carry signals both ways. One obvious application is in wire-guided weapon systems such as torpedos or missiles. Here one wishes to transmit information gathered by a detector at the weapon front end back to the control, and at the same time transmit commands from control to the weapon. In order to keep the connecting cable small and light, it is advantageous to use a single fiber for both purposes, and hence bidirectional couplers are called for. Interactive TV transmission is another application. Because of the large number of subscribers, using a separate fiber to transmit the interactive signals back to the station would greatly increase the number of fibers required; hence bidirectional transmission must be considered.

5.4 Isolators

The laser is like an RF generator, so it is obvious that one does not want any of the outgoing signal to be reflected back. Any such reflection would affect the laser cavity. Therefore the equivalent of a unidirectional resistor is required and garnets are being actively studied as a possible optical isolator.

Isolators will be particularly important in data buses, where there will be a large volume of signals flowing all over the place. This is an area with active research.

5.5 Spectral Filters

Spectral filters are required to deal with wavelength multiplexing. For window multiplexing, where there is typically several tenths of μm, separation between different channels, multilayer dielectric spectral filters are quite adequate. However, for wavelength multiplexing within the same window, gratings are needed to provide the necessary resolution.

There are many exciting ways of making effective spectral filters and a lot of physics research is going on in this area. Dr. Tien has discussed a comb-like structure etched on silicon which has functions similar to a grating. It is also possible to envisage a similar device by inducing dendrilitic growth on a crystalline substrate.

5.6 Measurement Equipment

There is an urgent need to standardize measurement methods and equipment. The progress is unfortunately very slow and fiber manufacturers are all too aware of the problem of correlating their own measurements with those made by other organizations. Measurement of a number of parameters needs to be standardized: attenuation coefficient, bandwidth, the numerical aperture relative to SiO_2, the fiber geometry and the cut-off wavelength for a single mode waveguide. The problem of standardization is aggravated by the fact that results depend on the launching condition of the signal and the way the fiber is wound.

5.7 Data Bus Technology

The devices discussed above become all the more important in the context of the data bus. For conventional point-to-point transmission, each link (be it fiber or copper cable) joins a transmitter at point A to a receiver at point B. A network would consist of a number of such point-to-point links in a delta

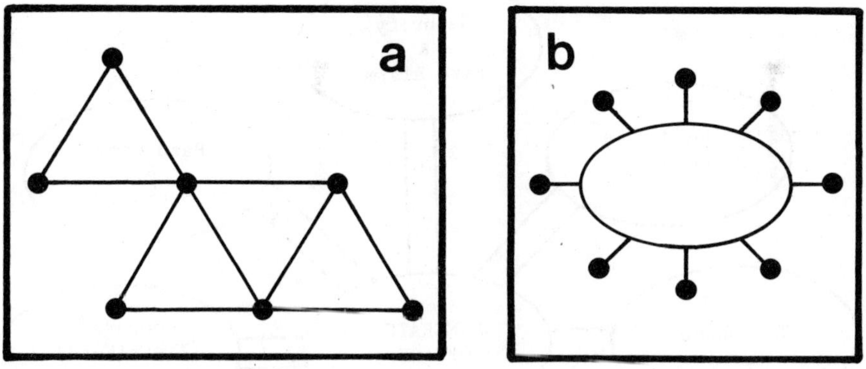

Figure 27: Networks connecting different points
(a) in a delta system (b) in a loop system.

arrangement as illustrated in Figure 27a. Such a network provides a certain amount of redundancy in that there are several routes linking any two points. However, by analogy with bus routes, it is clear that such a delta arrangement is not necessarily the most efficient. Now that a very high capacity transmission line is available, people are beginning to think about a new architecture to provide more efficient transport of data. A loop system (Figure 27b) comes to mind.

The importance of data bus systems for connecting many transmitters and receivers is increasing. To develop such systems, one must consider the problems of speed, error management, bus protocol and the interface with terminals and other networks.

Integrated office The office of the future will very probably integrate its various storage, processing and communication requirements into a network based on data bus technology (Figure 28).

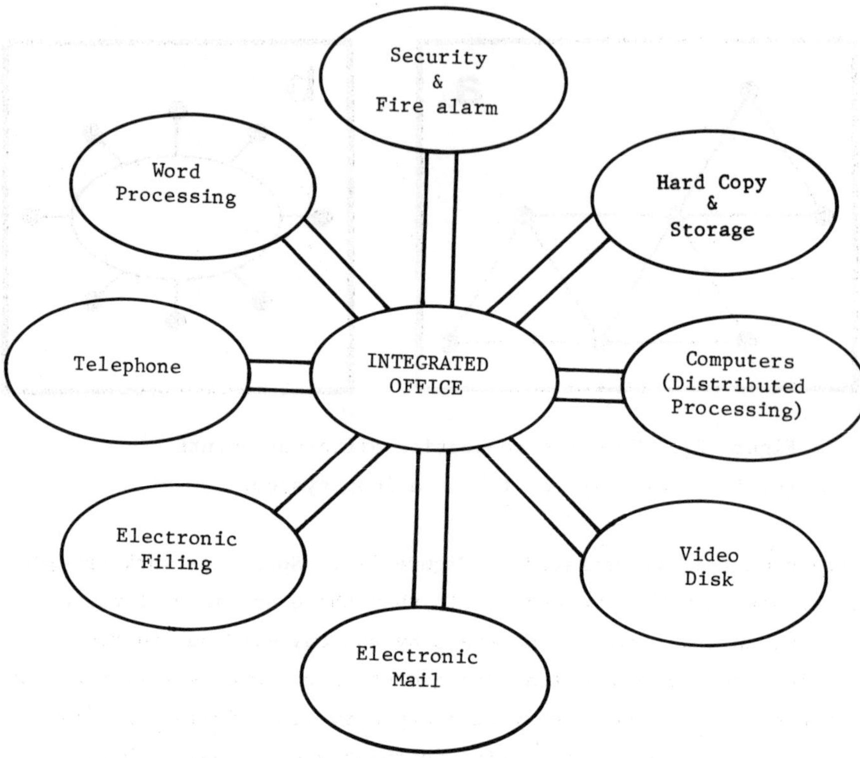

Figure 28: Integrated office.

Special application ITT recently built a special system for NASA (Figure 29). This system uses a 16 port star coupler with fiber optic bus interface and block synchronization to achieve 1 km data links between two computers (one for master control and the other for data management) and an archival mass memory with 10^{13} bits capacity. The system works at 100 Mb/s data rate.

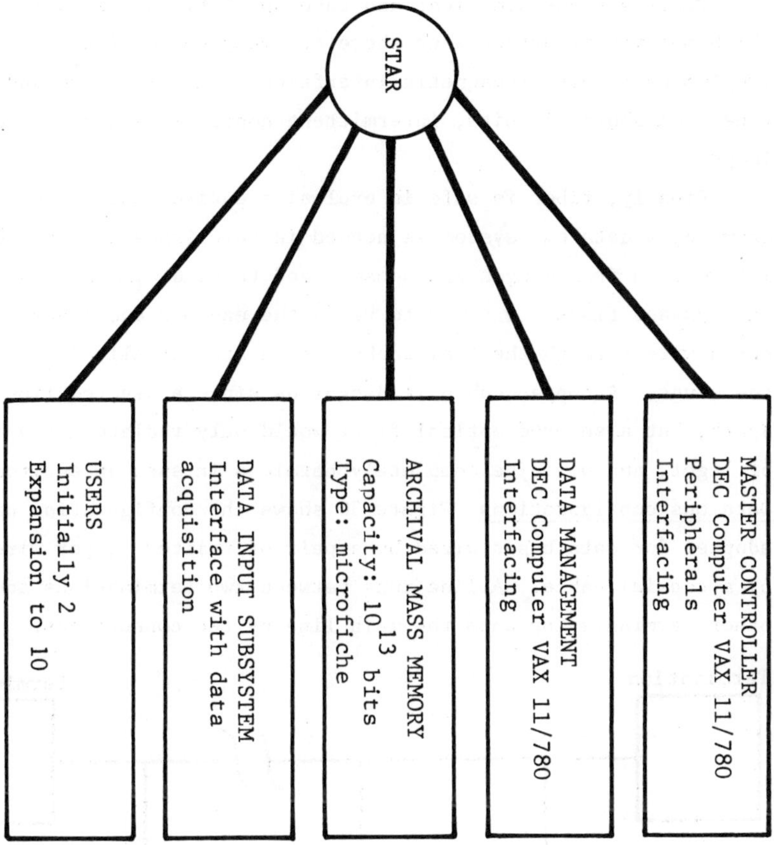

Figure 29: NASA data bus system.

<u>Why fiber optics</u> There are many reasons for preferring fiber optics to copper cables in implementing a data bus system. By far the most important consideration is the much higher data rate that can be achieved on an optical system. Moreover, fiber allows a larger number of terminals to be interconnected and the low loss permits larger terminal separation. For long distances, transit times become a limitation. For example, transit times over a few km of a few to tens of μs must be considered long when data are being transmitted at say 100 Mb/s. However, clever schemes have been devised to get around this.

Fiber systems are light and take up little space, a factor which becomes important when there are many connections. Fibers are immune to electromagnetic interference and lightning and are free from short circuits, intermittent connections and ground loops.

Finally, fiber is safe in explosive environments. For example, a data bus system is needed in aeroplanes to receive information from gauges and sensors and to send commands to engines and flaps. One way to reach the engines and flaps is to put a wire through the fuel tanks, which are normally situated in the wings. A copper cable is dangerous since a severed line may spark, but a severed optical fiber would only radiate a few mW of light, and would be completely harmless in such an environment.

Data bus configurations Figure 30 shows the configuration normally adopted for data buses wired by a pair of twisted copper wires or by a coaxial cable. A line runs between two terminations and the other terminals tap onto the main line by tee connections.

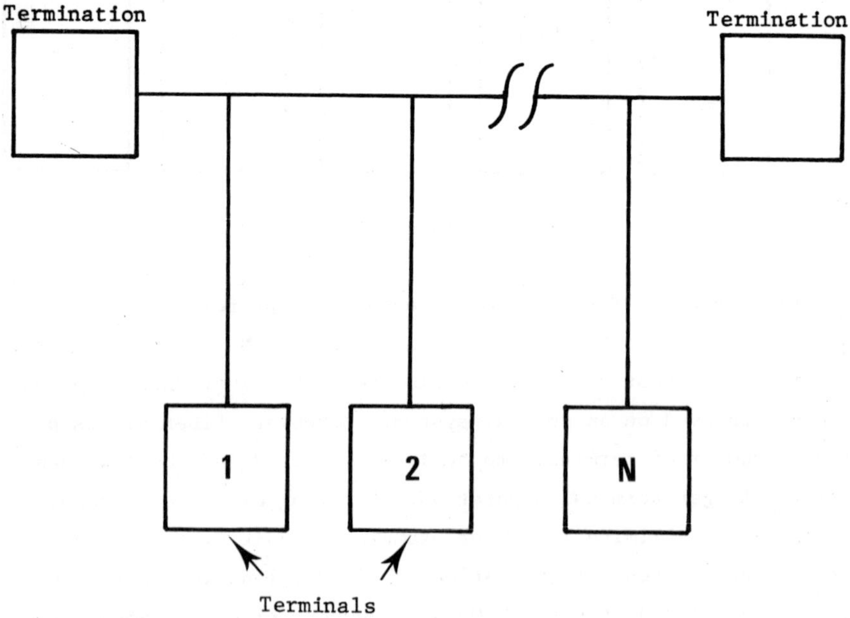

Figure 30: Configuration of data bus based on copper cables.

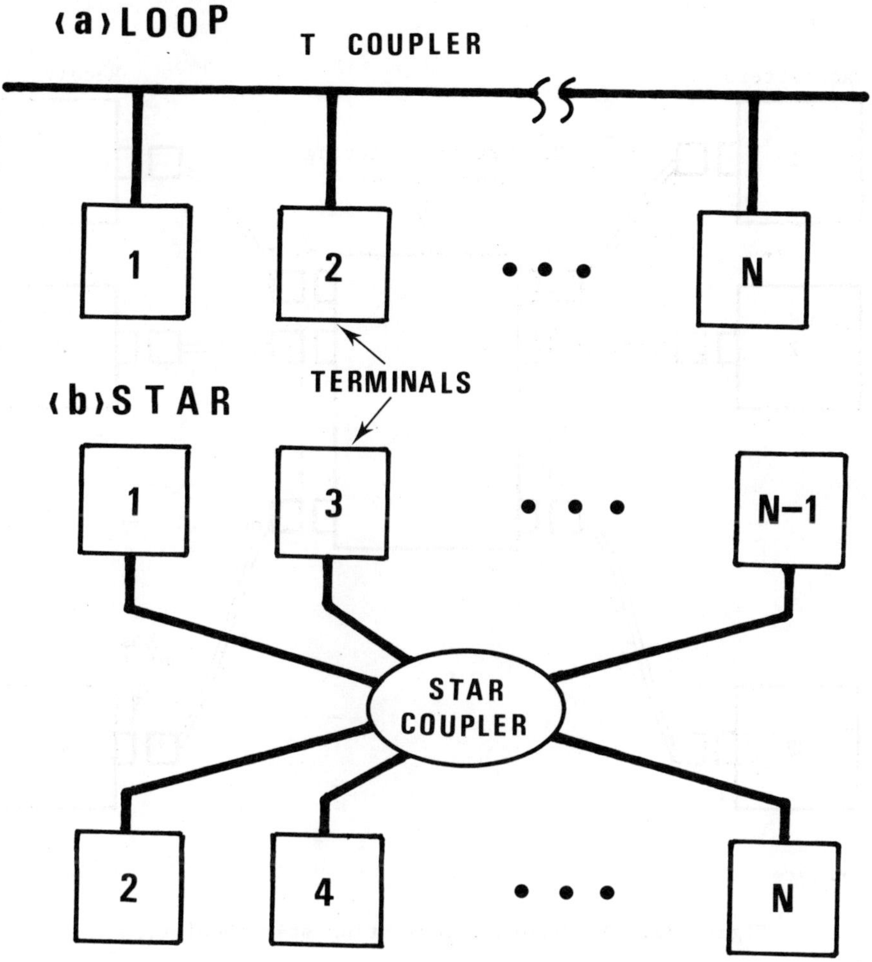

Figure 31: Optical data bus configurations

Optical systems can be wired in a similar manner using tee couplers tapping a loop (Figure 31a) or in a somewhat more **novel** scheme, using a star coupler (Figure 31b).

Figure 32 shows one 32 port transmitting star coupler that may be used for a data bus. This is a transmitting coupler in the sense that the 32 ports on one side are all transmitters, and the signal from each of these is coupled to each of the 32 ports on the receiving side. The insertion loss in a star coupler consists of two parts. There is first of all the inevitable

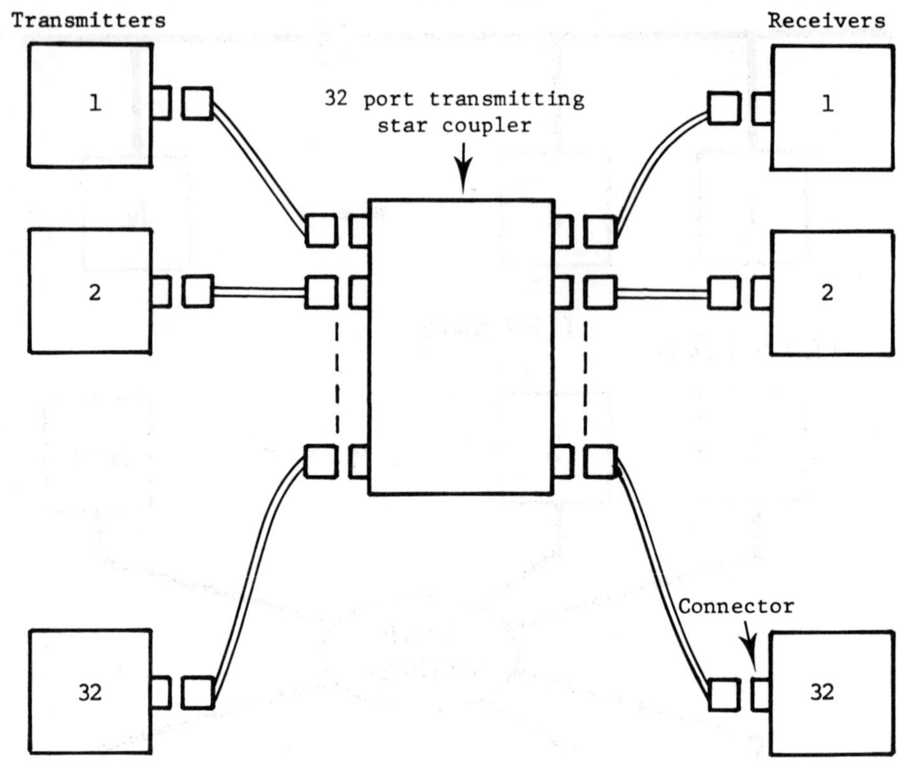

Figure 32: A 32-port transmitting star coupler.

division of power into each of the receiving ports, in this case resulting in a 1/32 loss factor. There are also excess losses over and above this; for a large number of ports this may be as large as 6 dB.

There are two possible ways of wiring onto a star coupler; these are illustrated in Figure 33 in the context of a reflecting star configuration. The transmitting and receiving signals for each remote terminal may be switched onto the same fiber by a bidirectional coupler (Figure 33a), or the two signals can be carried on two different fibers (Figure 33b).

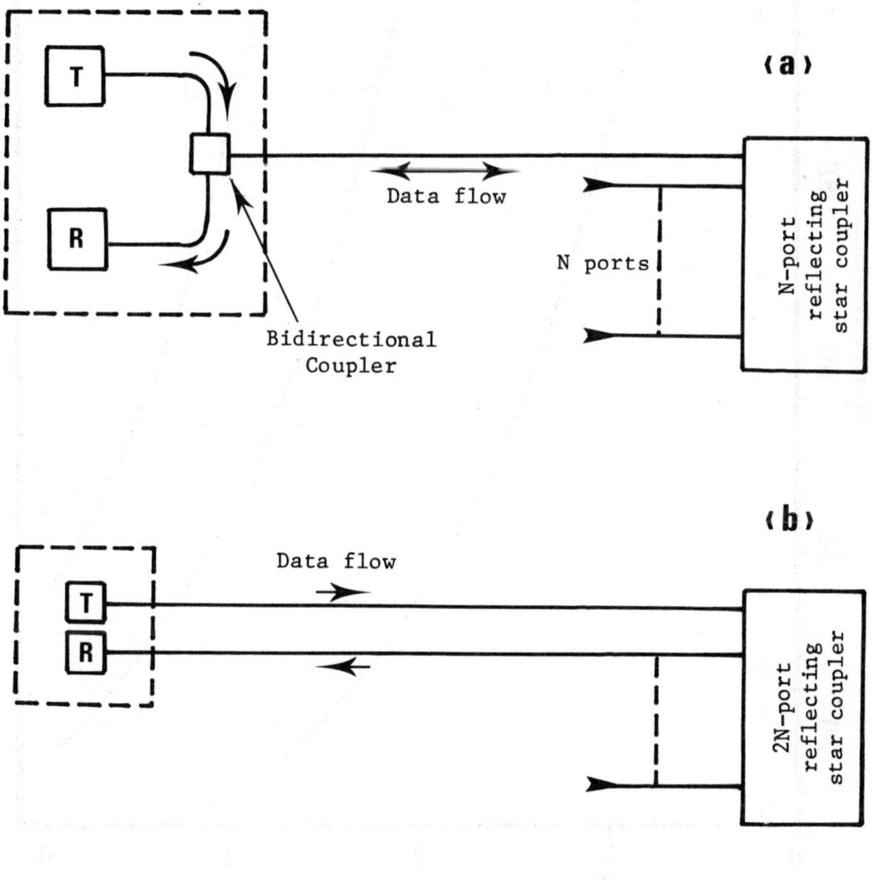

Figure 33: Two reflecting star configurations.

The number of terminals that can be connected in a star configuration is determined by the various losses. Two examples are shown in Figures 34 and 35. It is seen that 50-100 terminals are not difficult from the power budget point of view.

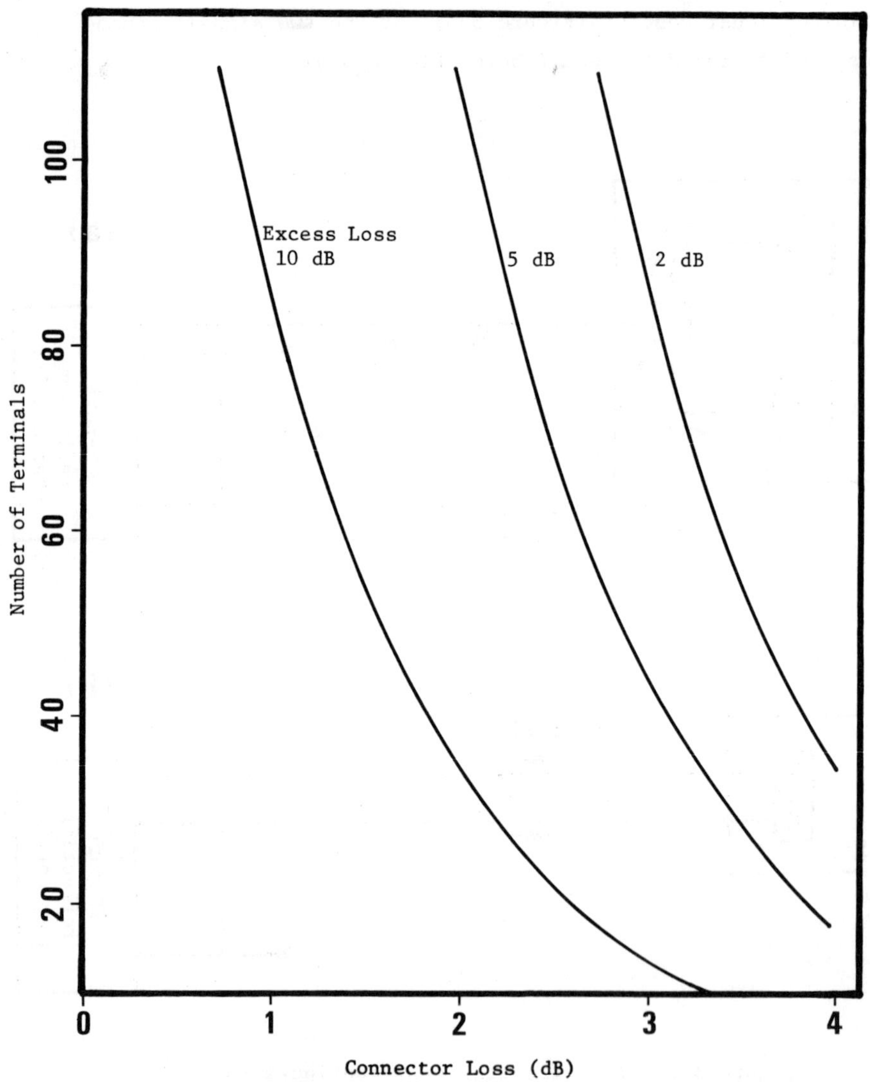

Figure 34: Maximum number of terminals in a transmitting star topology, assuming transmitter power = - 10 dBm, receiver sensitivity = - 45 dBm, loss between terminals = 0.2 dB.

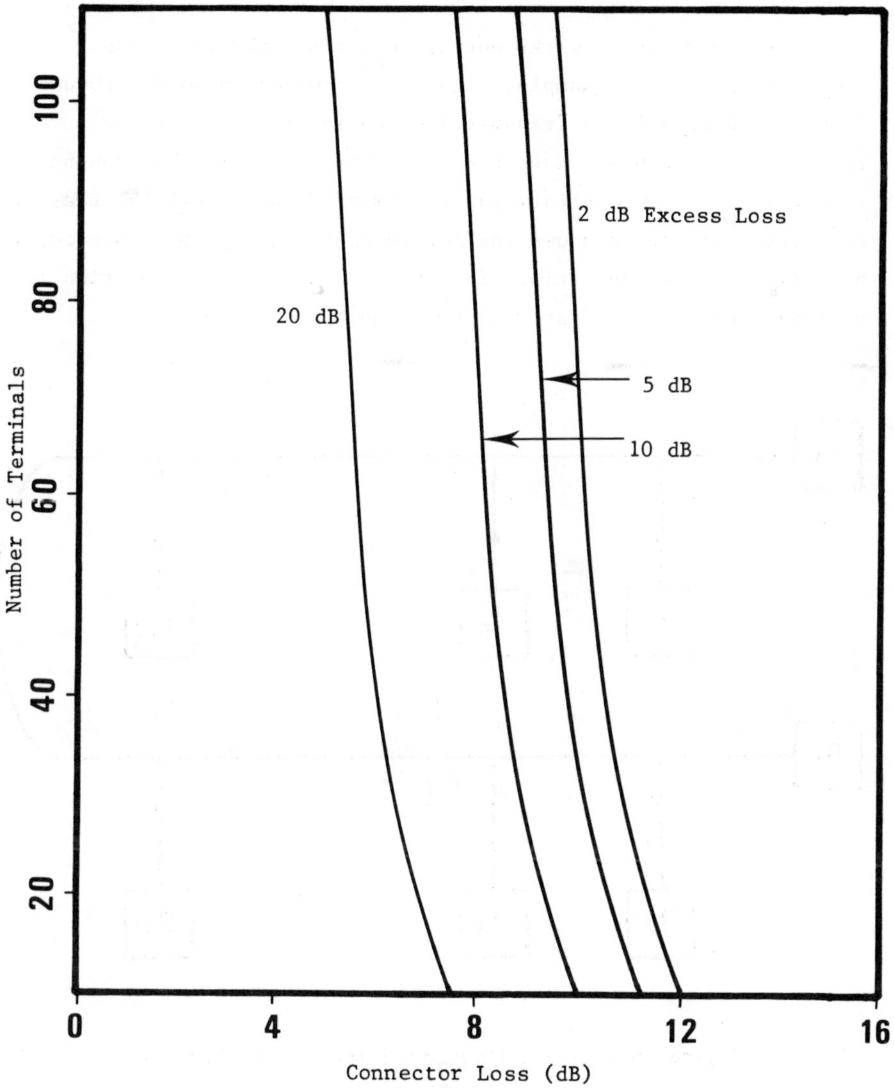

Figure 35: Maximum number of terminals in a transmitting star topology, assuming transmitter power = 0 dBm, receiver sensitivity = -65 dBm, loss between terminals = 2 dB.

The situation illustrated in Figure 35 corresponds to using a laser source. In this case the existence of many different paths makes modal noise a serious problem.

The power budget works out quite differently in a loop configuration. For example, Figure 36 shows an inbound/outbound loop, in which all the transmitters are in front of all the receivers. Figure 37 shows the number of terminals that can be supported, assuming uniform tap ratios of 3 dB for all the tee couplers. Figure 38 shows the improvement that can be achieved by tailoring the tap ratio. In any case, a loop configuration is suitable only for a relatively small number of terminals.

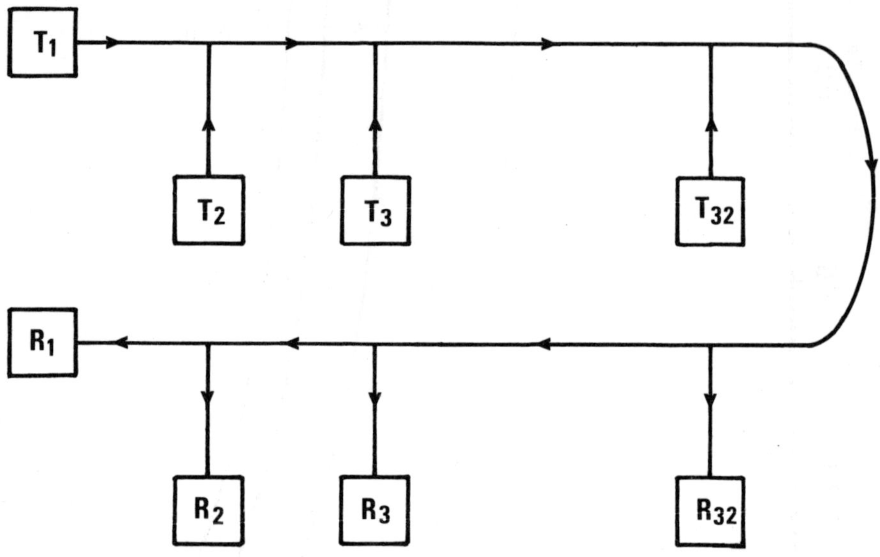

Figure 36: Inbound/outbound loop configuration.

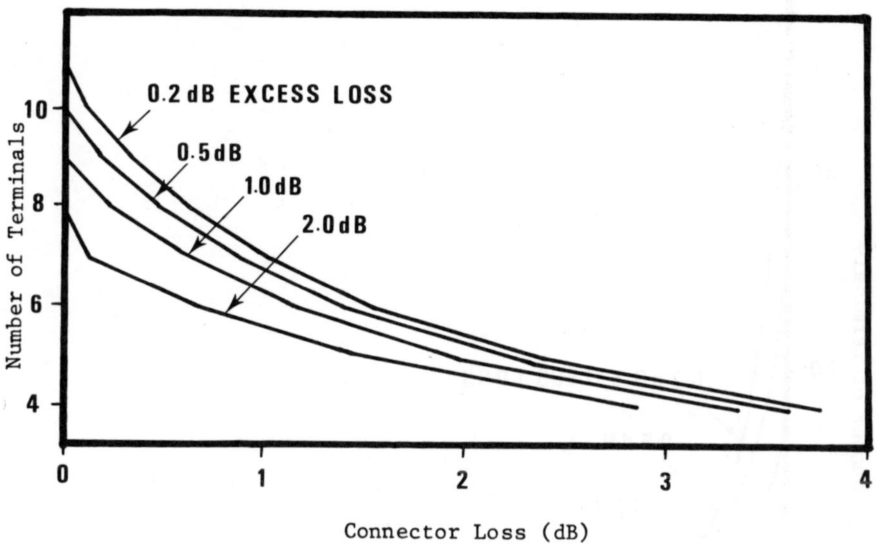

Figure 37: Maximum number of terminals in a loop topology, assuming transmitter power = 0 dBm, receiver sensitivity = -65 dBm, loss between terminals = 0.2 dB, tap ratio = 3 dB.

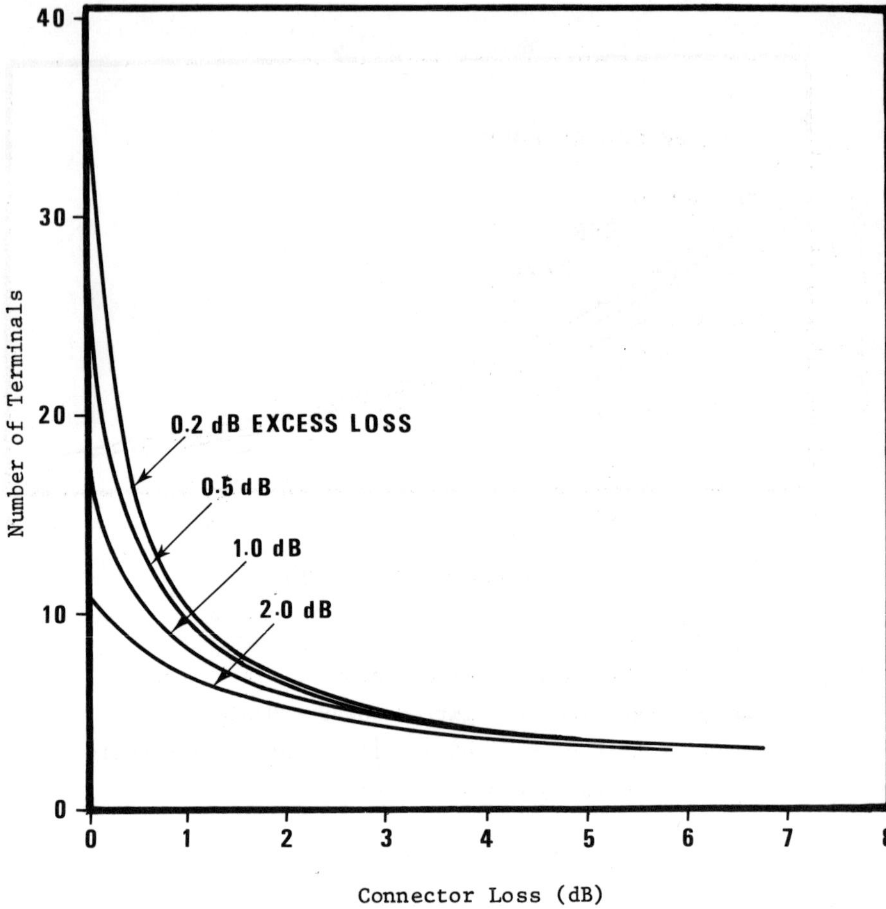

Figure 38: Maximum number of terminals in a loop topology. All parameters same as previous figure, except that tap ratio is tailored for each tee.

One can also have various combinations of such configurations. Figure 39 shows an example of a star-star configuration. Terminals A1, A2, ... are connected via short fibers (2 ft.) to star coupler A; B1, B2, ... are likewise connected to star coupler B; .... . Each coupler A, B, C is further connected to a central 3 port star coupler via long fibers (148 ft.). In this particular design, the path from A1 to A2 is short and goes through only one coupler, so the loss is only 19 dB. For A1 to B1, the loss is 54 dB. Since

A1 must receive and process signals from both A2 and B1, it needs to have a dynamic range of some 35 dB, which is extremely difficult to achieve in practice.

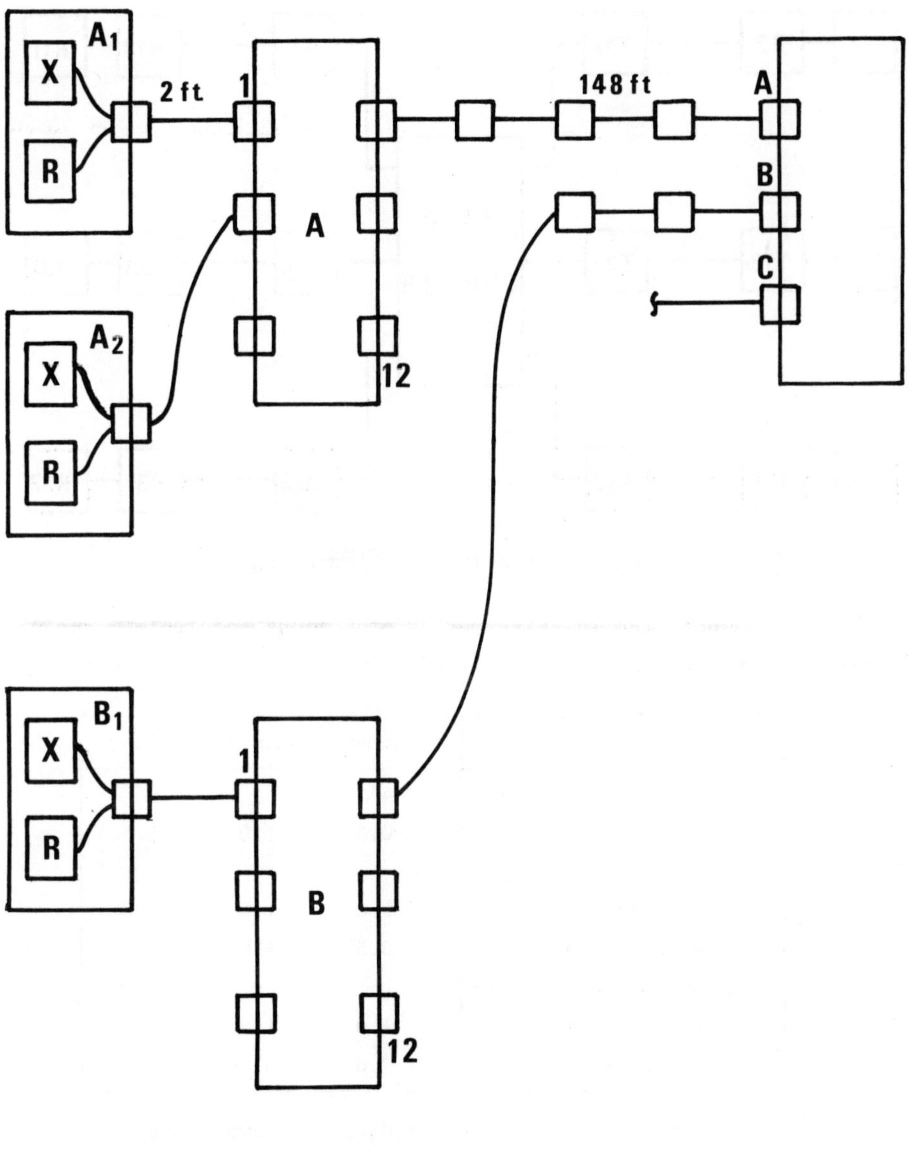

Figure 39: A star-star configuration.

Figure 40 shows a single star coupler together with a number of taps to achieve a star-loop topology.

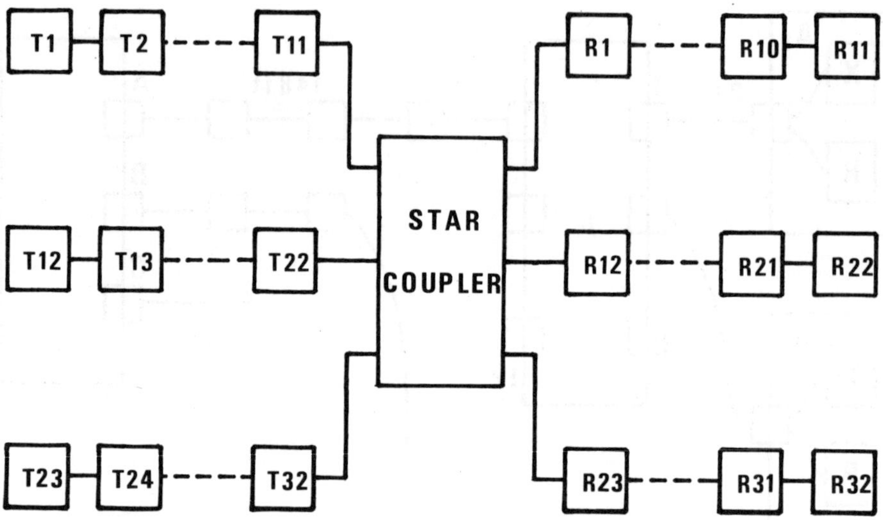

Figure 40: Star-loop configuration.

To evaluate the relative merits of various topologies, one has to assume certain losses. Typical values are as follows:

	Minimum	Nominal	Maximum
Fiber (db/km)	4	6	8
Connector (dB)	0.5	1.0	1.5
Tee coupler			
insertion (dB)	1.0	1.5	2.0
excess (dB)	0.5	1.0	1.0
Star coupler			
insertion (dB)	1.0	1.5	2.0
excess (dB)	2.0	3.0	4.0

Loss characteristics of optical data bus components.

A word should be said about the high loss assumed for the fibers. This is because data bus will probably use high numerical aperture (~ 0.3), large core (~100 μm) fibers, the loss of which is currently at about 6 dB/km. Based on the above loss parameters, one arrives at the following comparison between different configurations.

	Worst Loss (dB)	Best Loss (dB)	Dynamic Range (dB)	Optical Signal Range (dB)
Transmissive Star	41	19	22	12.5
Reflective Star	50	25	25	14
Star-Star	65	16	49	40
Inbound/Outbound Loop	311	-	-	-
Bidirectional Loop	186	-	-	-
Star/Loop	139	-	-	-
Loop/Star	42	15	27	23

Topology analysis summary.

From these figures, it is seen that the dynamic range requirement of the loop/star configuration (27 dB) is not excessive, and can be made to work in practice.

<u>Terminal architecture</u> In addition to connectors, directional couplers, transmitters and receivers, one needs a certain amount of electronics to handle the data, and to ensure that information is sent and received in an orderly fashion (Figure 41). First of all, analog signals must be converted to digital and coded in a certain format. Then there must be a certain amount of protocol, i.e. rules to specify when each terminal can transmit or receive. There must also be "handshaking" signals to establish that terminal X is transmitting to terminal Y. The data validation is to detect errors.

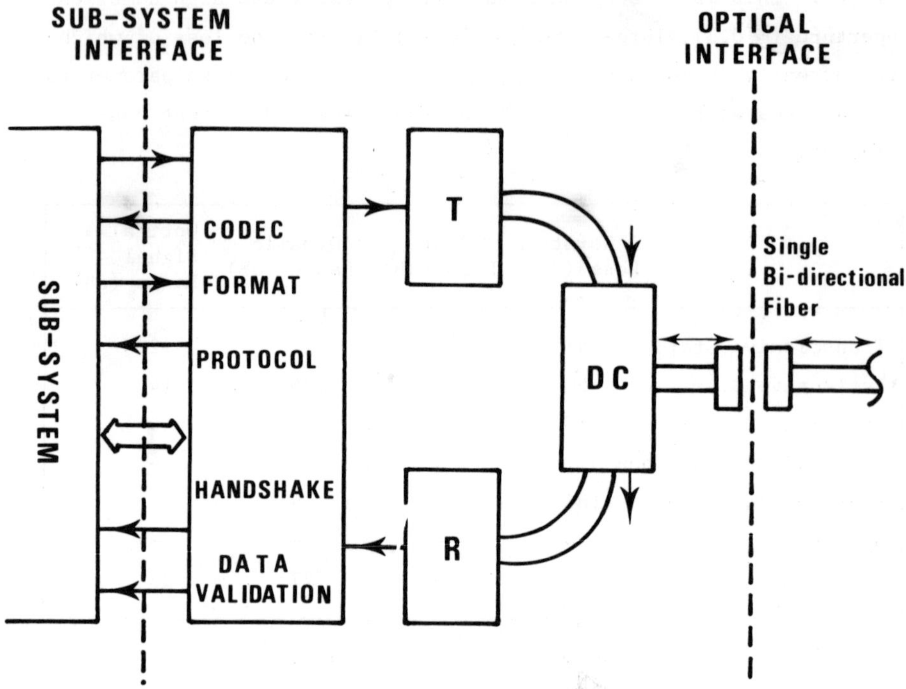

Figure 41: Terminal architecture.
T = transmitter, R = receiver, DC = directional coupler.

Time slot structure Figure 42 shows one possible time slot structure. It starts off with a guard length of 32 dummy bits to allow some time for the system to be brought up and for any transients to disappear. This is followed by synchronization, message type, destination and originator address, a spare 40 bits for contingency, error detection, and finally the message itself. It is seen that the overhead is rather high.

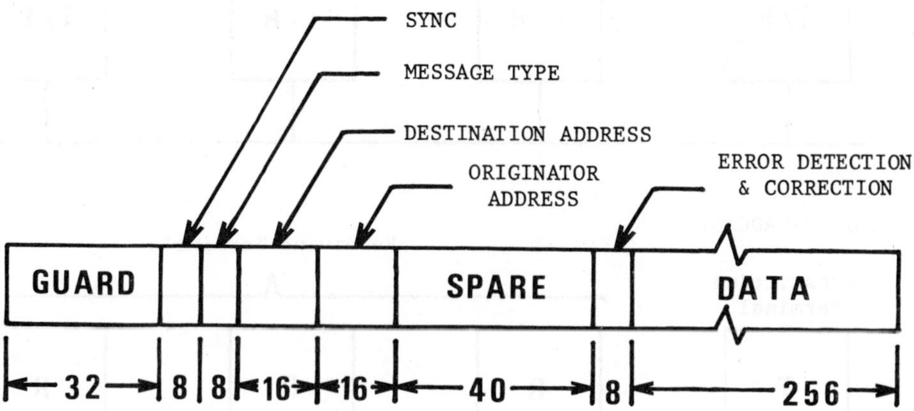

Figure 42: Time slot structure.

Control structure One can think of three ways of controlling a data bus. In the command/response scheme (Figure 43a), one special terminal acts as a bus controller and allocates transmission time to remote terminals upon request. The broadcast scheme (Figure 43b) is even simpler, in that there is only one transmitter, the other terminals all being receivers only. The autonomous system (Figure 43c) is more interesting. In this scheme, each terminal is self-sufficient and can originate a message and perform command functions as if it were the controller.

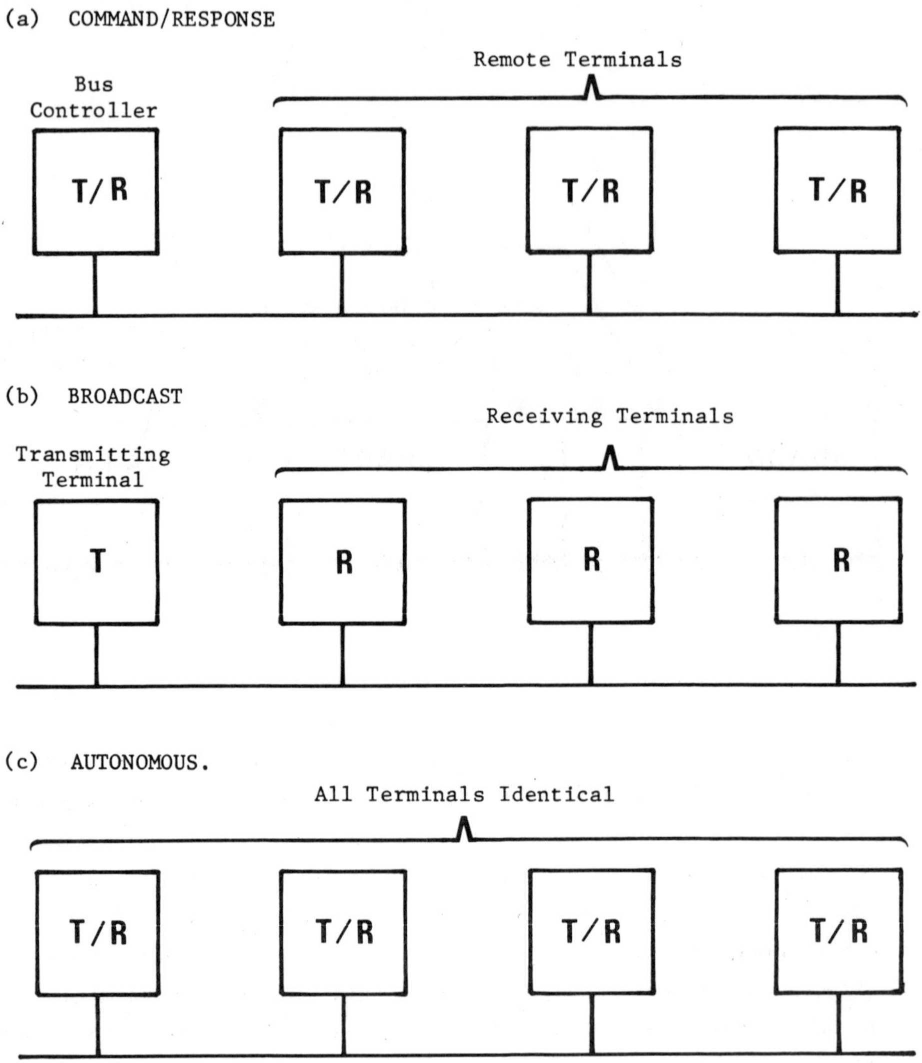

Figure 43: Data bus control schemes.

In order to achieve the autonomous scheme, the following ingredients are required: one must be able to receive the signal, address the gap counter in the receiver, count the minimum gap, and allow transmission to take place.

The absence of a single bus controller has certain advantages. First of all, the system is not vulnerable to failure at a single point, since any terminal can control the system. Secondly, the autonomous scheme is consistent with the trend towards distributed processing.

The autonomous system is also flexible in allowing addition or deletion of terminals. Moreover, addition of a new terminal would not normally require either hardware or softward change. The available bandwidth will be utilized more efficiently because one transaction can be set up while other transactions are under way. Each terminal can assess whether the bus is busy; if not, it can take up a large share or even all of the time. A second terminal coming on may request a portion of the time share. It is also possible to design protocol so that important messages can overide other transmission. All these features come under the heading of dynamical allocation of bandwidth.

5.8 Overview of Fiber Systems

In discussing associated components, as we have done in this lecture, it must not be forgotten that the entire fiber optics system is itself only part of the new information processing network which is becoming increasingly important. This network consists of three main components: computers, VLSI technology and fiber technology. Computers are increasing the capabilities in information analysis, processing and switching. VLSI technology is enabling devices to be available for handling increased capacity and signal speed. Fiber systems increase transmission capacity, and hence allow easier access to information resources. It is fair to say that these three form the cornerstones of the new information society. With rapid advances in all three sectors, we shall begin to harness information instead of being its slaves.

6. ECONOMICS OF OPTICAL FIBER SYSTEMS

The main thrust of this lecture is to suggest that the cost of fiber systems can be looked at in various different ways. Many people in comparing a fiber system to a copper cables system tend to look only at equipment costs. This is not necessarily the best approach, although this is often the only way since the immediate need of deciding how to deploy available capital in the purchase of equipment is usually the constraint. The more general aspects of the comparison are to be addressed.

6.1 General Economic Considerations

Since telephone networks form the principal application of optical fiber systems, it is profitable to take a look at how many telephones a country or region can support. It has been observed that the number of telephones per hundred population increases with GNP; in the US the figure is close to 100. Less developed countries may be at the level of say 25 telephones per 100 population. Examination of these figures then reveals the potential for market growth.

For countries with high GNP, further growth in demand may lie in video services instead of more telephones. For these countries, the number of subscribers who can afford video is really quite large, but for some reason people do not want video phones at the moment, and one might say that the social development lags behind technological progress. This example is meant to illustrate the fact that demand is not determined solely by price, but is affected by social conditions as well.

To understand and predict the impact of fiber systems, one must also distinguish the new market, in which fiber provides a hitherto unavailable service, from the replacement market, in which fiber replaces older forms of technology. (The same of course goes for any new product.) The economic considerations are quite different in these two cases. For example, in the case of telephone networks, fiber provides a replacement for conventional systems; in such cases the price differential has to be significant before a decision is likely to be forthcoming to adopt the new technology. (The situation may be somewhat different in less developed countries, where the number of telephones is not near saturation, so the anticipated growth provides a new rather than replacement market.) On the other hand, the tremendous bandwidth of optical fibers creates new markets, e.g. for systems which distribute large volumes of information to subscribers. The price that the customer is willing to pay in such circumstances may be quite different.

It is also necessary to distinguish between the early and mature phases of the market. For fiber systems, we are still in the early phase. Customers tend to anticipate improvements and greater future needs and usually want the best system available, even if the full performance of such a system may not be needed at the moment. This can create problems for the manufacturer, who obviously develops his products from the relatively simple to the more sophisticated. Yet he is often required to sell the latest and most sophisticated products first. In a settled, mature market the situation is quite different. The customer is likely to have clear objectives and specifications, and within these constraints, he will select a system almost entirely on the basis of price.

6.2 Basic Comparison

The simplest comparison that one can make is on an equipment by equipment basis, considering transmission medium, detector, associated electronics etc. separately. As we shall see, the comparison can be very different, depending on the job at hand.

Transmission medium One can envisage three different media: a twisted copper pair, coaxial cable, and fiber. The current costs per meter are as follows:

Medium	Cost per meter (US$)
Twisted copper pair	0.03
Coaxial cable	0.60 - 2.0
Fiber	0.30 - 0.60

But the relative merit depends on the application. If one wishes to transmit a large number of channels (e.g. 600 channel FDM telephone system) which can not be carried on a pair of copper wires, fiber is already competitive and in a good position to replace coaxial cables. For a small number of channels (e.g. 24 channel PCM), fiber still has some way to go before it can edge out the twisted copper pair. In other words the bandwidth is a cost factor.

The span length is another crucial factor. At distances of a km or so, a pair of copper wires can perform satisfactorily. But at say 10 km, a few repeaters may well be necessary, and the cost changes drastically. We shall see that the longer the span length, the more competitive fiber systems become.

Terminal cost In the present context, a terminal will mean, for copper systems, an amplifier for electrical signals and for an optical system, a device that converts an electrical signal to an optical output and vice versa. These would be needed both at the terminals and at the repeaters.

For a copper system, the cost of a repeater for 24-30 PCM channels is only tens of dollars. In an optical system requiring a laser source, the laser itself may cost $400 or more. However, if the system can be designed to work with an LED instead of a laser, the cost may be only $20-$30. The detector cost is still to be added.

Bandwidth Since bandwidth has a large influence on cost, it is necessary to distinguish between different fiber types. It is simplest to express various constraints in terms of the maximum distance that can be spanned without repeaters, as a function of

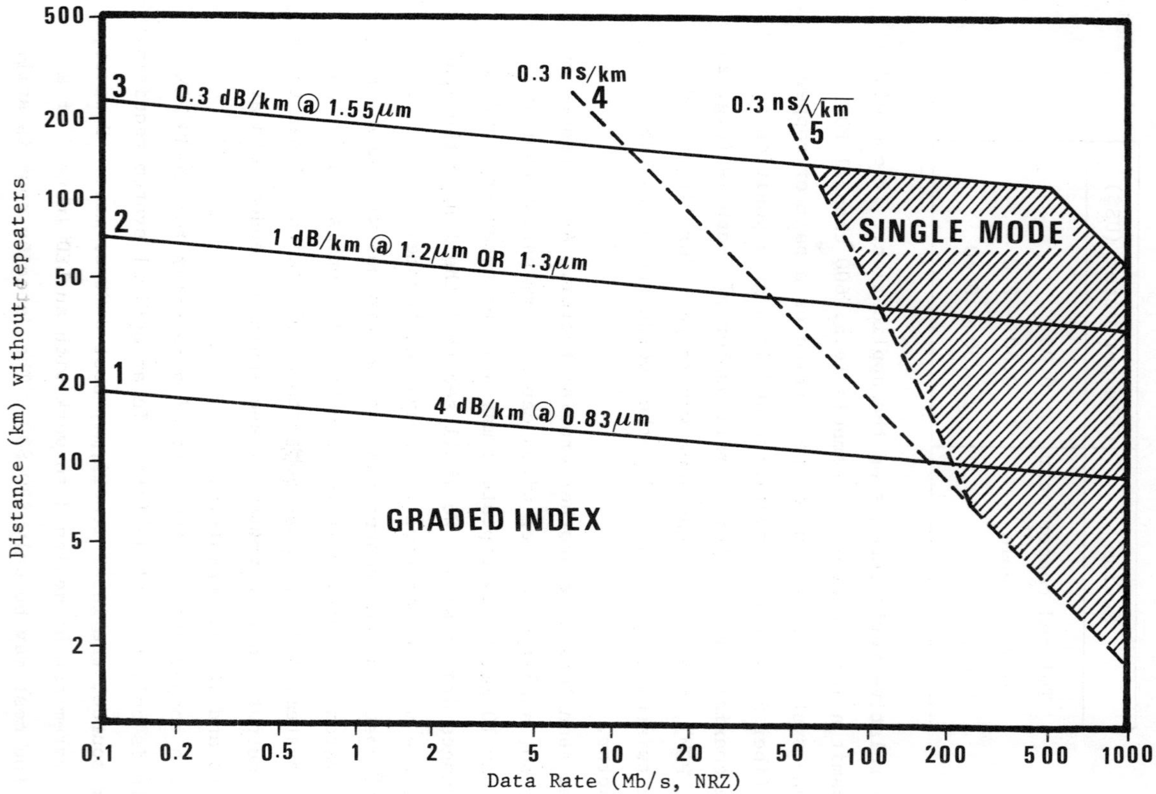

Figure 44: Limit on fiber systems.

data rate (Figure 44).

First of all, the loss in the fiber imposes a limit on the distance. As we have stressed, most of the loss is due to Rayleigh scattering by inhomogeneities, and goes as λ^{-4}. So as the operating wavelength is increased from 0.83 μm (curve 1), to 1.2 or 1.3 μm (curve 2) and 1.55 μm (curve 3), the loss drops from 4 dB/km to 1 dB/km and 0.3 dB/km. The distance spanned increases from about 10 km to 50 km to about 150 km.

These three curves show a slight dependence on data rate, basically for the following reason. The higher the data rate, the better must be the signal-to-noise ratio in order to ensure an acceptable error rate. Thus one needs greater received power and hence can achieve only a smaller span length.

Thus, for a given wavelength, loss considerations limit systems to the region below the corresponding curves (1, 2 or 3).

However, dispersion places another limit on the system. For a perfect step-index multimode fiber, the various modes propagate at different speeds and cause a differential delay time Δt over a length L given by

$$\Delta t \simeq AL$$

where A is a constant. Clearly Δt must be smaller than the time occupied by one bit of data, so

$$\Delta t \cdot R \ll 1$$

where R is the bit rate. So one arrives at the following constraint between data rate and span length:

$$RL \ll A^{-1}$$

For a typical value of A=0.3 ns/km, this condition restricts operations to the left of curve 4.

The theoretical bandwidth of a graded-index fiber is very high; in other words, such a fiber is designed to have a very small value of A. However, the presence of random imperfections turns out to imply another differential time delay due to intermode conversions:

$$\Delta t \simeq BL^{\frac{1}{2}}$$

(One may think of Δt as an error; the errors due to different parts of the fiber are statistically independent and hence are added together quadratically in the usual way for errors, resulting in an $L^{\frac{1}{2}}$ dependence.) The above condition implies a limitation

$$RL^{\frac{1}{2}} \ll B^{-1}$$

For a typical value of $B=0.3$ ns/$\sqrt{km}$, this condition restricts operation to the left of curve 5.

It is therefore seen that in the shaded area (long span length, large bit rate applications), only single mode fibers are adequate.

It should also be noted that there are other limitations due to the difficulty of modulating the laser at high rates.

Factors affecting cost When one discusses the cost per unit route length of a fiber system, two factors should be kept in mind. First of all, as system bandwidth increases, one needs more sophisticated electronics to deal with the modulation. At 1-20 MHz bandwidth, low frequency electronics is adequate. Above 20 MHz or so, medium frequency technology is required, so there is a quantum jump in cost per unit route length. At roughly 300 MHz, high frequency electronics is needed, with a further quantum jump in cost. The cost ratio between the high bandwidth and the low bandwidth can be a factor of ten. But notice that one is buying perhaps 100 times more capacity.

Secondly, for low system lengths (say 1 km or so), the low loss of the fiber is not being fully utilized, and the "overhead" associated with source, detector, etc. is being borne by a short route length; therefore the cost per unit length is high. As route length increases, cost per unit length drops. Beyond 100 km or so, the cost becomes dominated by the fiber plus repeaters if any, hence the cost per unit length approaches a constant.

To summarize the above discussion, it is convenient to consider the cost per unit bandwidth per unit distance, as a function of system bandwidth (Figure 45). For short systems, the electronics and peripherals rather than the transmission medium dominate, so there is little difference in costs between copper and cable systems. For long distances, the cost of the transmission medium becomes important, and the difference between copper and fiber shows up. The relative merit depends on bandwidth requirements. At small bandwidths, copper systems are still advantageous, but above say 30 MHz, fiber begins to win. At 1 GHz, the cost difference can be as much as a factor of ten.

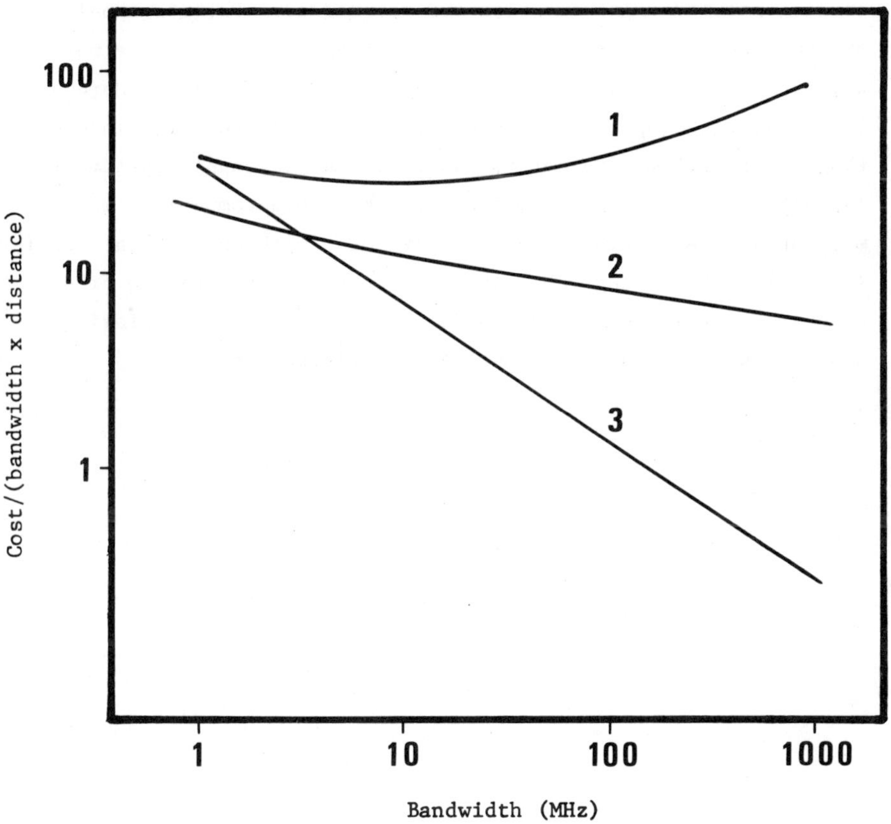

Figure 45: Cost vs system bandwidth
1 = short copper or fiber system
2 = long copper system
3 = long fiber system

Figure 46 shows the overall comparison in terms of present value annual cost (PVAC); such a comparison takes into account all relevant factors, e.g. cost of money. Figure 46a shows the relative PVAC against circuits increase per year, for two cases: 34 Mb/s and 140 Mb/s. In both cases, fiber is more economical. Figure 46b shows the relative PVAC for 160 circuits added per year against route length. For a 34 Mb/s system, fiber optics is more economical than coaxial cables. For 2 Mb/s, a copper pair is adequate and coaxial cables are not needed. For a new pair or new fiber, the cost is approximately the same and there may not be an overwhelming incentive to go to fiber optics. Moreover, if an existing copper pair is used, the cost is very low indeed; thus fiber systems are not yet able to replace copper systems for low data rates and short distances. The above comparison was made on the assumption of many fibers each with one optical carrier (spatial multiplexing), but it is now possible to use few fibers each transmitting on several carriers (wavelength multiplexing) and this would reduce the cost of long fiber systems.

(a)

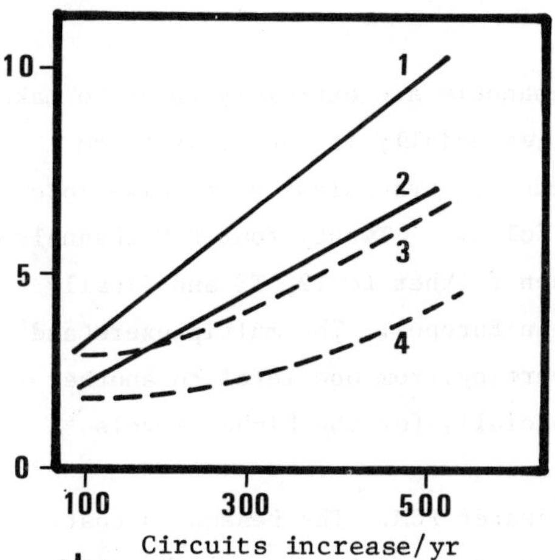

(a) Present value annual cost (PVAC) vs circuits increase per year

1 = 34 Mb/s coaxial
2 = 34 Mb/s fiber

3 = 140 Mb/s coaxial
4 = 140 Mb/s fiber

(b)

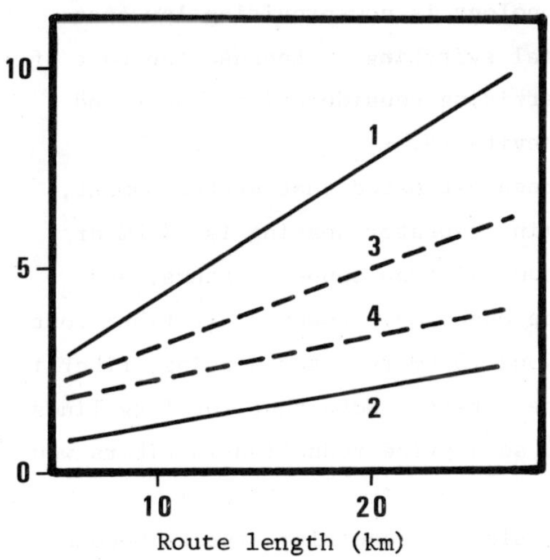

(b) Present value annual cost vs route length

1 = 2 Mb/s new pair or fiber
2 = 2 Mb/s existing pair

3 = 34 Mb/s coaxial
4 = 34 Mb/s fiber

Figure 46

6.3 Signal processing

Analog versus digital Analog channels are extremely cheap to make. Digital equipment is expensive, especially if one needs to do multiplexing. The general scheme for transmission of pulse code modulation (PCM) signals is as follows. Twenty four PCM channels are multiplexed to T1 level, then further to T2, T3 and finally T4 level at 274 Mb/s (280 Mb/s in Europe). The multiplexers and de-multiplexers needed for converting from one level to another are rather expensive items, especially for the higher levels (e.g. T3/T4 conversion).

Yet the current trend is towards PCM. The reason is that switching systems are a major cost component in a telephone network. The advance in IC technology is now providing low cost and powerful electronic digital switching. Because the cost of switching is very much an overriding consideration, the trend towards digital signals is inevitable.

For long lines, it has been estimated that at the moment, with fiber at 30¢/m and assuming repeater spacing is 15 km or more, then fiber is more economical than copper systems, and 1-2% market penetration can be expected. However, if fiber cost comes down to 10¢/m, then beyond 10 km repeater spacing, fiber is expected to capture 50% of the market. Note that for long lines the repeater costs are small, so a price reduction in fibers would have a large impact.

For systems below 10 km, signal processing aspects become dominant.

FDM versus TDM There is also the question of whether one should stay on frequency divisional multiplexing (FDM) or go to time divisional multiplexing (TDM). For telephone systems, it is clear that we shall go to TDM. But for TV transmission, the situation is different. At present, FDM is much cheaper for television. The reason is that the A/D conversion (say 6 MHz bandwidth analog to 100 Mb/s digital) is at the very edge of current technology, and hence quite expensive. FDM avoids such a conversion.

However, one disadvantage of FDM is that the required signal-to-noise ratio is quite high, so large span lengths are not possible.

Wavelength multiplexing Currently one works with only two wavelengths in two transmission windows, e.g. 0.85 μm and 1.3 μm. As we have said in a previous lecture, there is potential for further development in this direction. The economic attraction is that one can increase capacity without installing new fibers. However, one must bear in mind that there will be insertion losses for putting in each wavelength and these should be taken into account in the initial design.

6.4 Installation and Maintenance

Installation costs can be very high for copper systems and very low for fiber systems, but are not accurately predictable in general, since much depends on the particular situation. For example, if one happens to have a duct already installed, with sufficient space left for an additional fiber cable, then the cost would be minimal.

Fiber systems installed so far have been reported to be very reliable and maintenance free. This is because the distance between exchanges is often within fiber span length, so there is no field maintenance of repeaters. Under this situation the cost of maintenance is extremely low.

6.5 Right of Way

Fiber and copper systems are comparable in requiring right of way. In this respect, microwave systems have an edge and are the most severe competitor for fiber systems.

6.6 Network Planning

Network planning is a much more complicated exercise than simply comparing the cost of copper pairs versus fiber. All the factors discussed above (plus perhaps many other factors) must be taken into consideration. As an example, for countries which are building up a modern telecommunication industry, it is necessary to consider very carefully whether most components should be made domestically or whether certain parts should be purchased.

In conclusion, cost consideration for fiber systems should be carefully carried out in the most appropriate context. Otherwise the decision to adopt or reject the use of fiber system in a system network could be grossly incorrect.

INTEGRATED OPTICS AND INTEGRATION OF OPTICS WITH ELECTRONICS

P. K. Tien
Bell Telephone Laboratories
Holmdel, New Jersey 07733

ABSTRACT

Lately, research in integrated optics has entered into two different areas: (1) miniaturization and integration of optical devices for multiplexing, demultiplexing and signal processing at the optical level and (2) integration of optical devices with transistors for forming optoelectronic circuits. Both areas are very important in view of the needs in automation, sensors and optical communication. Since optical waveguides have always been the central focus of integrated optics, we devote the first two lectures to the concepts and mathematics of waveguides. Interesting topics, such as potential-well model of the waveguide and conformal transformation of the curve waveguides, will be discussed. We then turn to two-dimensional optics such as thin-film prisms and lenses, and refraction and reflection of light at tapered junctions. Here we deal with geometric optics in an "adiabatic" wave system. In the fourth lecture, we consider coupling between two waveguides or between two waveguide modes. It turns out that a simple coupled-mode theory can be applied to a large class of problems, including gratings, distributed feedback structures, directional couplers, as well as modulators and switches. In the final lecture, we discuss the difficult technology of optoelectronic integration and explore present state of the art and future trends of research.

AND

1. INTEGRATED OPTICS AND MODERN ELECTRONICS

1.1 Historical Review

Optics is an old science. In the past, since the invention of prisms and lenses several hundred years ago, most optical components were made of glass and they tended to be heavy and bulky. However, the availability of low-loss optical fibers in the early seventies brought optical communication into reality and there were sudden needs for compact optical devices and for integration of these devices into a reliable communication system. It is then obvious that the goals of integrated optics were set for miniaturization and integration of optical devices and systems. What we had in mind was medium or even large scale integration of optics, and such visions have definitely been inspiring in view of the success of microelectronics.

It was found that a thin layer of dielectric film (thin-film waveguide) or a strip of thin film (channel waveguide) deposited on a substrate is a perfect optical waveguide, provided that the refractive index of the film is larger than that of the substrate. In this arrangement, the high-index film which is sandwiched between the low-index substrate below and a low-index airspace above, acts as a lens. It has the capability of confining a light wave inside the film as well as guiding it along the film.

For the wavelengths of interest for optical communication (0.8 μm to 1.6 μm), the channel waveguide described above has a cross-section of typically 1 μm × 1 μm which is even smaller than the polysilicon lines presently used in LSI silicon circuits. The idea is to use these thin-film waveguides as basic structures of the optical devices and then to interconnect them by other thin-film waveguides to form optical circuits. For example, in lasers, the high-index film of the waveguide is a light-amplifying medium, and for modulators and switches, it is either an electromagnetic or electrooptic switching medium. The light wave is thus either amplified or switched by going through the waveguide. Naturally, the input and output terminals of the waveguide become the input and output terminals of the device. To form a laser-modulator circuit, it is a simple matter to connect the output waveguide terminal of the laser to the input waveguide terminal of the modulator. Thus, waveguides play a central role in integrated optics and for that reason, integrated optics is also called guided wave optics.

Therefore, we may miniaturize optical devices and circuits by using waveguides and thin-film technology. Moreover, by using III-V alloy semiconductor compounds such as GaAs, and InGaAsP, optical devices and transistors may be formed side by side on a common substrate. We soon expect a complete optical receiver or laser transmitter to be fabricated with all the electronics on a small chip of GaAs or InP. It is this new generation of opto-electronic circuits which fascinate us the most. They will have far-reaching consequences in automation and optical communication.

Because of the involvement of very difficult thin-film technology, active research in integrated optics was not begun until 1968. Even then, no more than a dozen scientists at Bell Laboratories and at the University of Washington were working on light wave propagation in thin-film waveguides. The subsequent invention of the prism coupler and m-line spectroscopy in 1969 greatly eased experiments on waveguide modes. In the following two years, the field mushroomed. When the first topical meeting on integrated optics was held in 1972 in Las Vegas, Nevada, it was attended by more than 150 scientists from 11 countries.

The accomplishments in the first five years were spectacular: thin-film prisms and lenses were invented in 1969. The possibility of forming these familiar optical components in thin film form attracted considerable attention in the scientific community. Another series of important inventions involved grating couplers (1970), distributed feedback lasers (1970), and use of gratings in waveguides for Bragg reflectors and optical filters (1969-1972). Tapered-film couplers were first reported in 1971, and by 1973, one began to consider use of tapered thin-films for forming light-guiding interconnections among devices. An integrated optical circuit involving a glass waveguide coupled to a silicon detector was reported in 1973.

Between 1973 and 1976, scientists began to concentrate their efforts on optical modulators, switches and beam deflectors. Magneto-optic waveguides were formed by growing iron garnet films on gallium garnet substrates, and electrooptic waveguides, by diffusing titanium into lithium niobate substrates. Some of the switches were formed in the form of a directional coupler. To ease tolerances for the fabrication of directional couplers, a reversal delta-beta arrangement for the electrodes was invented. In addition, light-beam deflectors involving the interaction between an acoustic wave and a light beam in a waveguide was studied extensively.

Lasers and detectors used for optical communication involve either AlGaAs alloy compounds grown epitaxially on GaAs substrates or InGaAsP alloy compounds grown on InP substrates. We will discuss these materials more in detail later. One realized at that time that all sorts of optical devices, passive or active, may be made from either of the two above alloy compounds, and integrated optical circuits formed in that way are called monolithic circuits. Use of semiconductor technology for optical devices and systems are welcomed by the electronics industry. It turns out that the above alloy compounds are also better materials for transistors because of their high mobilities and saturation velocities. In fact, the fastest transistor amplifiers are made from GaAs. The possibility of forming optical devices and transistors on a common substrate allows us to combine optics with electronics in integrated circuits, thus opening up new possibilities and new applications to modern electronics.

1.2 Integrated Optics and Silicon Technology

In the last section, we have shown that research in integrated optics has entered into two different areas: (1) integration of optical devices to form an optical circuit, and (2) integration of optical devices with transistors to form an optoelectronic circuit. Interesting optical circuits being investigated are multiplexing and demultiplexing circuits for optical communication. These circuits will likely be made in titanium diffused lithium niobate containing optical filters, channel switching network and branching waveguides. Obvious examples of optoelectronic circuits are integrated optical receivers, transmitters, and repeaters. In most cases, it is convenient to convert optical signals into electric pulses and process the signal at the electrical level. In fact, we have to do this, if the integrated optical circuit is to be used in a multimode fiber system. Single mode waveguides used in the optical circuit will not accept light from a multimode fiber, though it can deliver light to the multimode fiber fully without sacrificing any performance. In that sense, optoelectronic circuits are important in that signals may be in optical form or electrical form, or both.

It is interesting to compare the highly successful silicon technology with the technology which we expect to achieve in integrated optics. In spite of more than ten years of phenomenal growth, silicon technology still advances rapidly. The design-rule was 6 μm in 1978 and it is about 2.2 μm now. The number of devices per chip still increases at a rapid pace; it doubles about every two years. One example of this is the random access memory. We recall that a mere four years ago, large semiconductor manufacturers were competing with one another for the first 64 K RAM. In the last year, 256 K RAM was announced by Bell Telephone Laboratories and the same capability will soon be available at Motorola, IBM, Hitachi and Fujitsu.

However, we may soon witness fundamental limits of silicon electronics. Conservatively, we place a limit on design rule at 1 μm for silicon circuits, beyond which little can be gained. Although transistors have been fabricated in our laboratory with a gate length as short as 700 A, the limit in current density that a metal line can carry due to metal migration will come much sooner. In fact, this limit has already been reached in the present 2.2 μm design-rule for aluminum.

With the one micron design-rule in silicon technology in mind, the RC constants (resistance-capacitance product) of metal, polysilicon, and diffusion lines are respectively 10 picoseconds (ps), 8 nanoseconds (ns) and 6 ns for a path one millimeter long. The delay in a pass-over transistor is about 40 ps and that of a basic inverter pair, 200 ps. In a typical circuit design, one uses metal lines to distribute dc voltage or current, and polysilicon and diffusion lines to carry signals. It is easy to see that the speed of the silicon circuit is limited by the RC constants of the polysilicon and diffusion lines and not the time constant of the transistor itself. Hence, one will not gain much by making devices smaller than one-micron design rule. It is thus difficult to build a silicon digital circuit with a speed faster than 10 ns (100 Mb/s). One can gain at least a factor of 10 in speed by using metal lines for signal and logic, but that requires three levels of metal, which are again very difficult. The present 256 K RAM has an access time of 100 ns and a 100 Mb/s analog-to-digital converter has been demonstrated.

In one-micron design-rule, metal lines are 1 μm wide and polysilicon and diffusion lines are 2 μm wide. A basic inverter occupies an area of 25x10 μm. The optical devices used in integrated optics are much larger. One can build a laser in an area of 200x40 μm and a directional coupler, an area of 750x60 μm. However, the sizes of optical devices come into circuit performances in a different way. The RC constants in electrical circuit correspond to the material and waveguide dispersions of the optics. The dispersions are extremely small, much smaller than 1 ps/mm and may be neglected. The bandwidths of optical devices are thus virtually unlimited at the optical level.

Pure optical logic has not been considered in integrated optics. Modulators and switches discussed in integrated optics are optoelectronic devices in which, an applied electrical signal causes the phase, or polarization, or direction of propagation of a light wave to vary. In those devices, sizes of the optical devices determine sizes of the electrodes which introduce RC constants in the electrical circuits. For example, a directional coupler with electrodes 750 μm x 30 μm area has a RC constant of 100 ps which is much smaller than those of the polysilicon and diffusion lines in silicon circuits discussed earlier. It is thus safe to say that the speeds of the opto-electric circuits are limited by the speeds of electronics. Since either GaAs or InGaAs will be used for optoelectronic circuits one gains two accounts from these materials: (1) mobilities and saturation velocities are higher in GaAs and InGaAs than in silicon, and (2) semi-insulating substrates are available for these materials which cut down parasites. We believe that optoelectronic circuits can be built with the speed on the order of 1 to 10 Gb/s. Performances of various technologies are compared in Figure 1.

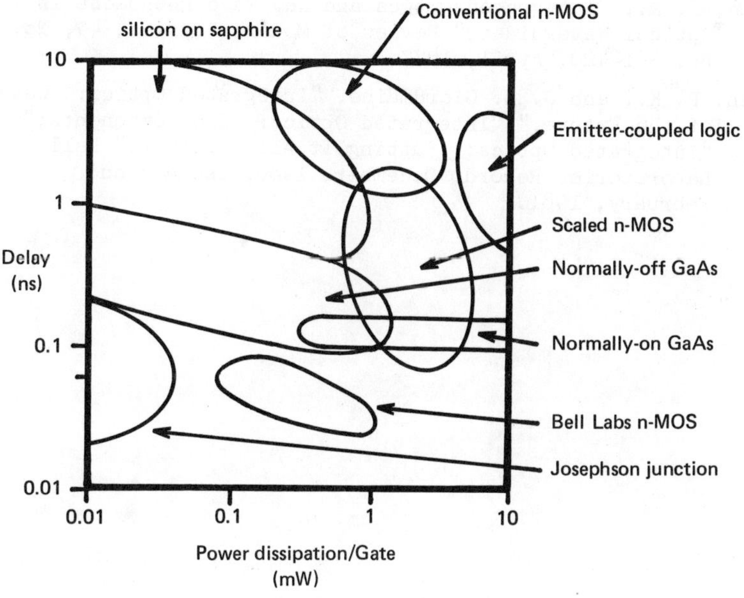

Figure 1. Circuit performance of various technologies

REFERENCES

1. Kogelnik, H., "Limits of Integrated Optics," Proceedings of IEEE, 69, No. 2, February 1981.
2. Mead, C., and L. Conway, "VLSI Systems," Chapters I, II, III, IV, Addison-Wesley, 1980.
3. Tien, P. K., "Integrated Optics and New Wave Phenomena in Optical Waveguides," Review of Modern Physics, 49, No. 2, pp. 361-420, April, 1977.
4. Tien, P. K., and J. A. Giordmaine, "Integrated Optics: Wave of the Future;" "Integrated Optics: The Components;" "Integrated Optics: Putting It All Together," Bell Laboratories Record, December, 1980, January and February, 1981.

2. DIFFUSED AND CURVED WAVEGUIDES

2.1 Introduction

In the late 60's, the concept of dielectric waveguides was well understood from the microwave theory. Still, experiments on light wave propagation in a thin layer of dielectric film was a scientific adventure. It was then a great satisfaction that waveguide modes can be displayed as m-lines by a prism coupler as discussed in the last lecture. In the meantime, important theories were written for a better understanding of guided wave optics. These include calculation of field distribution in a channel waveguide by Marcatili (1969), formulation of radiation and waveguide modes by Marcuse (1969), and the zigzag wave model of the slab waveguides by Tien (1970).

The zigzag wave model provides a clear picture of wave propagation and has been used extensively in device analyses. However, rigorous solutions and field distributions of the waveguides have to be derived from Maxwell's equations. In going through these computations, the reader should not be confused by the complexity of the field components involved. For that purpose, a unified theory is developed which is applied first to slab waveguides, then channel waveguides, and finally, diffused and curved waveguides as the problem becomes increasingly complex. The material used here draws heavily from the Ph.D. thesis of Dr. A. S. H. Liao.

2.2 Wave Equation and the Plane-Wave Solution

The central point of discussion in this lecture is wave equation and its solutions for various types of waveguides. Consider a wave equation in an isotropic medium of refractive index n. Denote $k = \omega/c$, where ω is the angular frequency and c is the velocity of light in vacuum. We have then

$$\left(\frac{\partial^2}{\partial x^2} + \frac{\partial^2}{\partial y^2} + \frac{\partial^2}{\partial z^2}\right) E = -k^2 n^2 E \tag{1}$$

The simplest solution of the above equation is a plane wave

$$E = A \exp(i\omega t \mp ik_x x \mp ik_y y \mp ik_z z) \tag{2}$$

where

$$n\bar{k} = k_x \hat{x} + k_y \hat{y} + k_z \hat{z}$$

and $\hat{x}$, $\hat{y}$, $\hat{z}$ are unit vectors in x, y, and z directions, respectively. Substitute (2) in (1), we obtain immediately,

$$k_x^2 + k_y^2 + k_z^2 = k^2 n^2 \tag{3a}$$

If k_x or K_y or k_z is imaginary, we write $k_i = -ip_i$. We have therefore three other forms of (3). They are

$$-p_x^2 + k_y^2 + k_z^2 = k^2 n^2 \tag{3b}$$

$$k_x^2 - p_y^2 + k_z^2 = k^2 n^2 \tag{3c}$$

$$k_x^2 + k_y^2 - p_z^2 = k^2 n^2 \tag{3d}$$

Equations (2) and (3) are of prime importance, and we shall discuss them in detail. First of all, $n\bar{k}$ is called a wave vector. It specifies the direction of wave propagation, $\bar{k}/k$. It also specifies frequency ω, or wavelength in the medium λ/n, since $nk=\omega n/c=2\pi n/\lambda$. Next, k_x, k_y, and k_z are components of the wave vector, and as shown in (3), they add vectorally to the total wave vector $n\bar{k}$. For that reason, they are called wave constants. For example, if k_x is real, the solution of the wave equation in (2) has functional dependences in x in the form of $\exp(i\omega t+ik_x x)$ which are sinusoidal functions. On the other hand, if k_x is imaginary and is equal to $-ip_x$ as specified in (3a), the wave solution has the functional dependences, $\exp(-p_x x)$ or $\exp(+p_x x)$, which are exponentially decreasing and increasing functions. Hence, as a plane wave propagates in a particular direction, the wave may appear as an exponential function in one direction and sinusoidal functions in other two directions as indicated in (3b), (3c), (3d), or it may appear as sinusoidal functions in all three directions as indicated in (3a). We also notice that if p_x is positive, the wave decreases exponentially in the positive x direction and because of this, p_x is called loss constant.

The plane wave shown in (2) is very important in our analysis, since waves in a rectangular waveguide can be expressed as a superposition of plane waves. We have already shown earlier in the zigzag-wave model that one plane wave propagates toward lower-right and the other toward upper-right. As the wave propagates in a diagonal direction, the wave motion viewed in the y-direction, for example, is represented by the wave vector, $k_y \hat{y}$. The wave appears having a wavelength equal to $2\pi/k_y$, and a wave velocity ω/k_y. This velocity is always larger than the velocity, c/n and is called phase velocity of the wave.

With the understanding of the plane wave solution of wave equation described above, we proceed to develop a unified theory for analyses of slab, channel, diffused and curved waveguides.

2.3 Field Components and TE and TM-Like Waves

The field components of the light wave in a waveguide may be computed from two curl equations of the Maxwell equations which have been used before to derive the wave equation (1). The curl equations are:

$$\bar{\nabla} \times \bar{E} = -\mu \frac{\partial}{\partial t} H$$

$$\nabla \times \bar{H} = n^2 \varepsilon \frac{\partial}{\partial t} E \quad ; \quad c = 1/(\mu\varepsilon)^{\frac{1}{2}} \; ; \; k = \omega/c \tag{4}$$

where, for example,

$$\bar{\nabla} \times \bar{E} = \begin{vmatrix} \hat{x} & \hat{y} & \hat{z} \\ \frac{\partial}{\partial x} & \frac{\partial}{\partial y} & \frac{\partial}{\partial z} \\ E_x & E_y & E_z \end{vmatrix}$$

Here we use MKS units and, as usual, μ and $n^2\varepsilon$ are permeability and dielectric constant of the medium. We assume that the medium is free from charges and currents. If there were currents in the medium, the refractive index, n, would be complex. Such will be the case when a semiconductor laser medium is considered.

Consider specifically a slab waveguide in Figure 2a or a channel waveguide in Figure 2b in which, a light wave guided by the waveguide propagates from $z=-\infty$ to $z=+\infty$ in the form of $\exp(i\omega t - i\beta z)$. Here, according to the notation used in the literature, we set $k_z=\beta$. The two curl equations in (4) can be converted into four scalar equations in terms of six field components E_x, E_y, E_z, H_x, and H_z. Since we only have four equations, it is convenient to express E_x, E_y, H_x, and H_y in terms of E_z and H_z. The result is

$$E_x = -i \frac{1}{\eta^2 k^2 - \beta^2} [\, \beta \frac{\partial}{\partial x} E_z + \omega \mu \frac{\partial}{\partial y} H_z \,]$$

$$E_y = -i \frac{1}{n^2k^2-\beta^2} [\, \beta \frac{\partial}{\partial y} E_z - \omega \mu \frac{\partial}{\partial x} H_z \,]$$

$$H_x = -i \frac{1}{n^2k^2-\beta^2} [\, \beta \frac{\partial}{\partial x} H_z - \omega\varepsilon n^2 \frac{\partial}{\partial y} E_z] \qquad (5)$$

$$H_y = -i \frac{1}{n^2k^2-\beta^2} [\, \beta \frac{\partial}{\partial y} H_z + \omega\varepsilon n^2 \frac{\partial}{\partial x} E_z]$$

First of all, we have derived the above equations assuming that a light wave propagates in the z direction. We call it a Z-system. We recall that in the last lecture, x was the direction of light wave propagation, which by the same token, may be considered as a X-system. It is possible to convert equations in the Z-system to those of the X-system by simply replacing x by y, y by z and z by x, because of the cyclic nature of the curl equations.

Here is the key to the solution: If we take $\partial/\partial y >> \partial/\partial x$, we find in (5) that E_x and H_y depend on H_z only, and E_y and H_x depend on E_z only. The above condition is certainly true in slab waveguides, since there, $\partial/\partial x = 0$ (note that we are in the Z-system). We have thus two sets of independent solutions for the slab waveguides. The first set involves E_x, H_y and H_z only and is called TE waves since it does not involve E_z. The second set involves E_y, H_x and E_z only and is called TM waves since it does not involve H_z. Even in the channel waveguide where the condition, $\partial/\partial x=0$, is not true. We still can set up two independent solutions for the waveguide and consider E_x, H_y and H_z as the principal field components for the TE-like waves and E_y, H_x and E_z as the principal field components for the TM-like waves. We will show in detail later how to handle this kind of situation.

2.4 Mode Equation of a Slab Waveguide

In a slab waveguide shown in Figure 2a, it is easy to see that the fields have to be sinusoidal in region I and exponentially decaying to infinity in regions II and III. Therefore, the waves are in the form of (3a) in region I and in the form of (3c) in regions II and III. Consider a TE wave consisting of field components E_x, H_y and H_z only. Since $\partial/\partial x = 0$ in a slab waveguide, it is convenient to take in region I,

$$H_{z1} = A_1 \sin (k_y\, y) \tag{6}$$

We do not know at this moment where the axis y=0 should locate. We could assume that the top boundary of region I is at $y=W_A$, and the lower boundary at $y= -W_B$ where $W_A + W_B = W$ which is the thickness of the light-guiding layer. The values of W_A and W_B can be determined later by matching boundary conditions. The advantage of doing this is The optics involved in the region above y=0 and that involved in the region below y=0 should be similar in nature. Once we know the wave solution for the region above y=0; we should be able to write a similar solution for the region below y=0 by inspection.

In region III, we take

$$H_{z3} = A_3 \exp\, [-p_{y3}\, (y-W_A)] \tag{7}$$

Substituting (6) and (7) into (5) and drop terms involving $\partial/\partial x$, we have

$$E_{x1} = -i\, \frac{1}{k^2 n_1{}^2 - \beta^2}\, [\, \omega\mu k_y\, A_1\, \cos(k_y\, y)\,]\ ;$$

$$E_{x3} = -i\, \frac{1}{k^2 n_3{}^2 - \beta^2}\, [-\, \omega\mu p_{y3}\, A_3\, \exp\{-p_{y3}\, (y-W_A)\}\,]$$

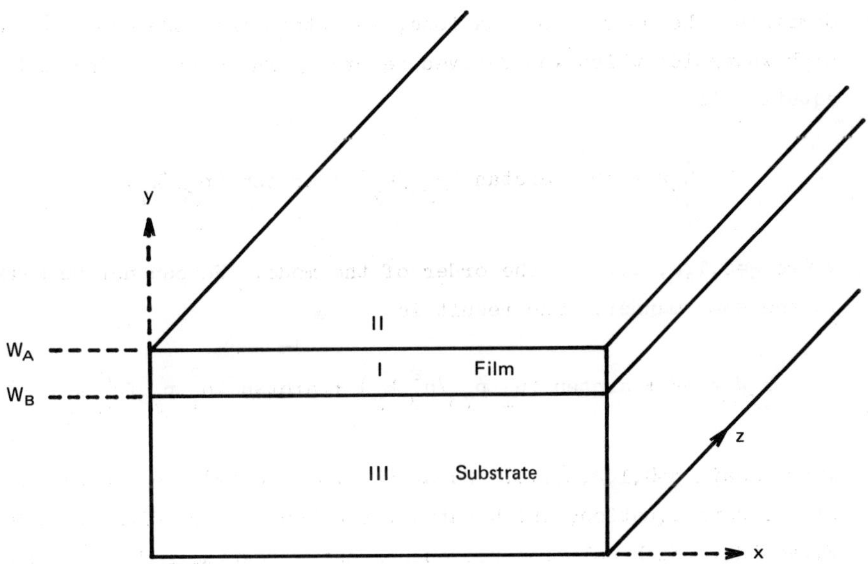

Figure 2a. Slab Waveguide

Equating E_{x1} and H_{z1} in region I to E_{x3} and H_{z3} in region III at the interface $y=W_A$, and using (3) we obtain immediately

$$\tan (k_y W_A) = p_{y3}/k_y$$

Similarly we may write by inspection,

$$\tan (k_y W_B) = p_{y2}/k_y$$

Combining the above two equations, we obtain the mode equation for the slab waveguide which was derived before by ray optics. The mode equation is

$$k_y W = q\pi + \arctan (p_{y3}/k_y) + \arctan (p_{y2}/k_y) \tag{8}$$

where q=0,1,2,3,....is the order of the mode. We can derive a TM wave in the same manner. The result is

$$k_y W = p\pi + \arctan (n_1^2\, p_{y3}/n_3^2\, k_y) + \arctan (n_1^2\, p_{y2}/n_2^2\, k_y) \tag{9}$$

where again p=0,1,2,3,.... Equations (8) and (9) are the eigensolution of the wave equation, which satisfy the boundary conditions of a slab waveguide. Each solution represents one waveguide mode. In the zigzag wave model we have shown that the largest possible value of k_y is $k(n_1^2 - n_3^2)^{1/2}$ if $n_3 \geq n_2$, and is $k(n_1^2 - n_2^2)^{1/2}$ if $n_2 \geq n_3$. Therefore by assuming $n_3 > n_2$, $q\pi < kW(n_1^2 - n_3^2)^{1/2}$ for both (8) and (9). The total number of modes that a slab waveguide can carry including TE and TM waves is less than $2kW(n_1^2 - n_3^2)^{1/2}/\pi$ in which, we notice that $(n_1^2 - n_3^2)^{1/2}$ is the numerical aperture and $kW(n_1^2 - n_3^2)^{1/2}$ is a V-number.

2.5 Mode Equation of a Channel Waveguide

In the channel waveguide shown in Figure 2b, the condition $\partial/\partial y \gg$

$\partial/\partial x$ is not strictly true. The field varies both in y and x, and for that we take in region I,

$$H_{z1} = A_1 \sin(k_x x) \sin(k_y y) \tag{10}$$

In this case, we do not know how to locate the point x=0, y=0. As before, we could assume the top, bottom, left and right boundaries of the region I be $y=W_A$, $y=-W_B$, $x=-X_B$, and $x=X_A$, where $W=W_A + W_B$ and $X=X_A + X_B$. Following the method used by Marcatili, we consider fields in the regions I, II, III, IV and V only and ignore those at the four corners of the waveguide. From (10), we can write by inspection

$$\begin{aligned} H_{z3} &= A_3 \sin(k_x x) \exp[-p_{y3}(y-W_A)] \; ; \\ H_{z2} &= A_2 \sin(k_x x) \exp[-p_{y2}(y+W_B)] \; ; \\ H_{z4} &= A_4 \sin(k_y y) \exp[-p_{x4}(x-X_A)] \; ; \\ H_{z5} &= A_5 \sin(k_y y) \exp[-p_{x5}(x+X_B)] \; , \end{aligned} \tag{11}$$

for regions 3, 2, 5 and 4, respectively. Of course, k_x, k_y and β in region 1, k_x, p_y and β in regions 2 and 3, and p_x, k_y and β in regions 4 and 5, should satisfy (3) noting that $k_z = \beta$. We have therefore 9 unknowns: E_{z1}/H_{z1}, A_3/A_1, A_2/A_1, A_4/A_1, A_5/A_1, k_x, k_y, W_A and X_A. There are only 8 boundary conditions involving tangential electric and magnetic fields at $y=W_A$, $y=-W_B$, $x=X_A$ and $x=-X_B$. It is then possible to freely choose one of the unknowns, say E_{z1}/H_{z1}, so that the solution of the wave equation will come out in the simple possible form.

We could choose either a TE-like or a TM-like wave in the waveguide. Taking the TE-like wave as an example, E_x, H_y and H_z are major field components. It is interesting that if we look at y-direction, regions 1, 2 and 3 form a slab waveguide. On the other hand, if we

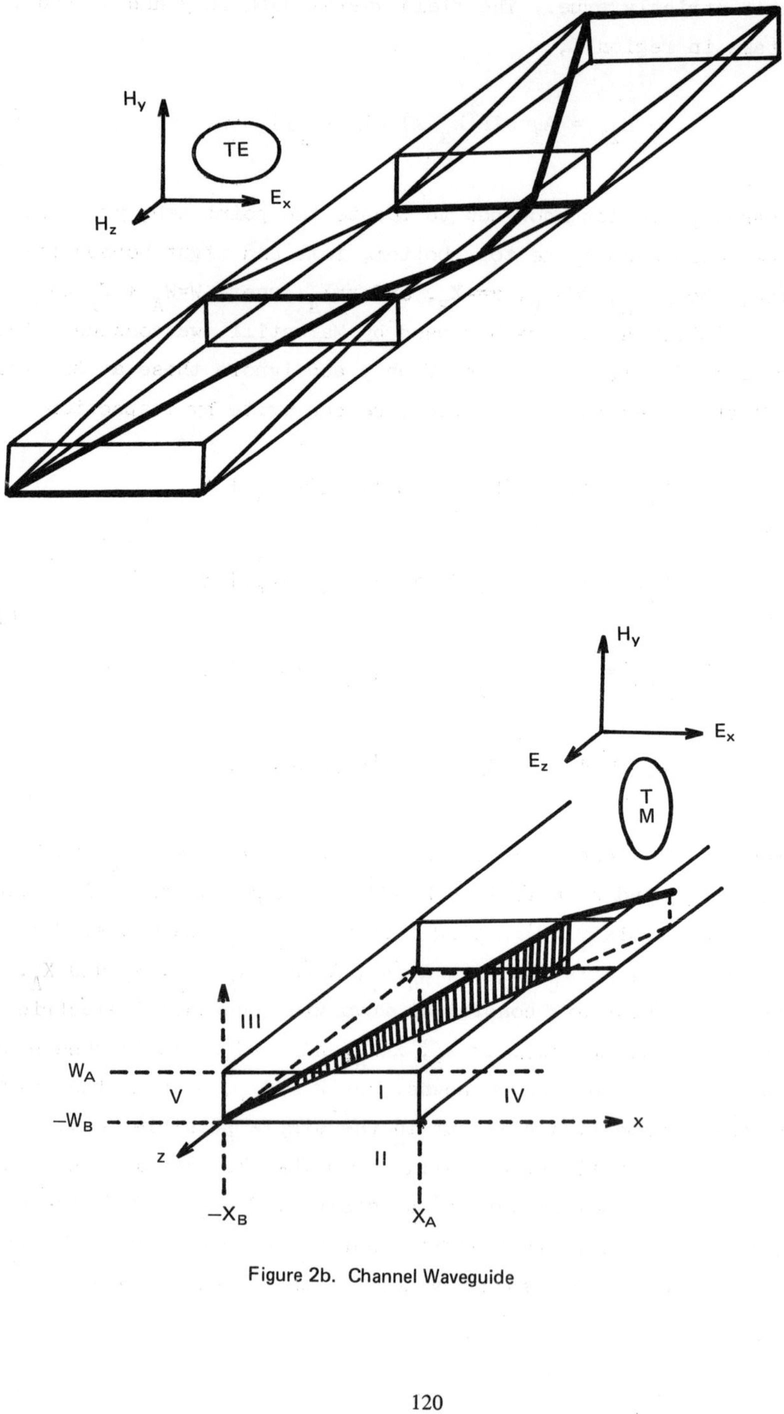

Figure 2b. Channel Waveguide

look at x-direction, regions 1, 4 and 5 form another slab waveguide. We call the former, slab-waveguide-Y, and the latter, slab-waveguide-X. Let us add another field component E_z to E_x, H_y and H_z chosen earlier. Among these fields, E_x, H_y and H_z form a TE wave in slab-waveguide-Y and H_y, E_x and E_z form a TM wave in slab-waveguide-X. The slab-waveguide-Y will then provide a mode equation similar to (8) which determine k_y and the slab-waveguide-X will provide a mode equation similar to (9) which determine k_x. The two mode equations completely characterize a channel waveguide provided that we can ignore E_z in slab-waveguide-Y and H_z in slab-waveguide-X. All seem to be straightforward.

Unfortunately, once we have E_z and H_z, there will be E_y and H_x from (5). E_y, H_x and H_z will introduce a TE wave in slab-waveguide-X and E_y, H_x and E_z, a TM wave in slab-waveguide-Y, in addition to the waves we already chose for the waveguides. If that were the case, the result would be a mixed TE-TM wave which we chose not to have in the beginning.

To avoid this difficulty, we select the ratio E_{z1}/H_{z1} such that $E_{y1}=0$. With this choice, there will not be a complete set of the field components for a TE wave in slab-waveguide-X or for a TM wave in slab-waveguide-Y. More importantly, we set the Poynting vectors $E_{y1} x H_{x1}\ \hat{z}$ and $E_{y1} x H_{z1}\ \hat{x}$ equal to zero. The power flow due to these components no longer exists. Indeed, we shall show that such a choice yields a solution which is TE-like in slab-waveguide-Y and TM-like in slab-waveguide-X or vice-versa. Such solutions are the basic solutions of a channel waveguide. Any solution which has a mixed TE-TM wave may be considered as a superposition of these basic solutions.

Now that $E_y=0$ in region 1 and thus in all other regions, by substituting (10) and (11) into (5), we have:

$$E_{z1} = -\frac{\omega\mu k_x}{\beta k_y} A_1 \cos(k_x x)\cos(k_y y)$$

$$E_{z2,3} = \frac{\omega\mu k_x}{\beta p_{y2,3}} A_{2,3} \cos(k_x x)\ \exp[-p_{y2,3}(|y|-W_{B,A})]$$

$$E_{x1} = -i \frac{\omega\mu}{k_y} A_1 \sin(k_x x) \cos(k_y y)$$

$$E_{x2,3} = i \frac{\omega\mu}{p_{y2,3}} A_{2,3} \sin(k_x x) \exp[-p_{y2,3}(|y|-W_{B,A})] \qquad (12)$$

By equating H_z and E_x in region 1 to those in regions 2 and 3 at $y=-W_B$ and $y=W_A$, respectively, we obtain a mode equation identical to (8) which is,

$$k_y W = q\pi + \arctan(p_{y3}/k_y) + \arctan(p_{y2}/k_y) \qquad (13)$$

Similarly,

$$E_{z1} = -\frac{\omega\mu k_x}{\beta k_y} A_1 \cos(k_x x) \cos(k_y y)$$

$$E_{z4,5} = \frac{\omega\mu p_{x4,5}}{\beta k_y} A_{4,5} \cos k_y y \exp[-p_{x4,5}(|x|-X_{A,B})]$$

$$H_{y1} = -i \frac{(n_1^2 k^2 - k_y^2)}{\beta k_y} A_1 \sin(k_x x) \cos(k_y y)$$

$$H_{y4,5} = -i \frac{(n_{4,5}^2 k^2 - k_y 2)}{\beta k_y} A_{4,5} \cos(k_y y) \exp[-p_{x4,5}(|x|-X_{A,B})] \qquad (14)$$

Again, by equating E_z and H_y in region 1 to those in regions 4 and 5 at $X=X_A$ and $X=-X_B$, respectively, we obtain another mode equation,

$$k_x X = p\pi + \arctan [(n_1^2 k^2 - k_y^2) p_4 / (n_4^2 k^2 - k_y^2) k_x]$$
$$+ \arctan [(n_1^2 k^2 - k_y^2) p_5 / (n_5^2 k^2 - k_y^2) / k_x] \qquad (15)$$

which is similar to (9), where p=0,1,2,3....and q=0,1,2,3. For each set of values of p and q, there is a waveguide mode in the channel waveguide. Equations (13) and (15) completely characterize the mode TE_{qp}. It is easy to see that the maximum number of the modes that a channel waveguide can carry is less than $2k^2WX(n_1^2-n_3^2)^{\frac{1}{2}}(n_1^2-n_4^2)^{\frac{1}{2}}/\pi^2$, if $n_3 \geqslant n_2$ and $n_4 \geqslant n_5$.

2.6 Effective Index Method

We shall now describe the effective index method which is a very powerful method and has been used extensively for solving ridge waveguides and various laser structures. The method stems from two-dimensional optics to be discussed in the next lecture. We may write the wave equation (1) in the form:

$$\left(\frac{\partial^2}{\partial y^2} + \frac{\partial^2}{\partial z^2}\right) H_z = -(k^2 n_i^2 - k_x^2) H_z = -k^2 n_{i-eff}^2 H_z \qquad (16)$$

where $-k_x^2$ have replaced $\frac{\partial^2}{\partial x^2}$ and i=1,2 and 3 for regions 1,2 and 3, respectively. We have now converted a three-dimensional wave equation into a two-dimensional equation in which k_x does not appear explicitly. In a way, we are entitled to do so, since k_x is identical in all the three regions considered. If we use (16) to describe the wave in a channel waveguide, the channel waveguide becomes a slab waveguide composed of regions 1, 2 and 3 which have effective refractive indices as

$$n_{1-eff}^2 k^2 = n_1^2 k^2 - k_x^2 \; ;$$

$$n_{2-eff}^2 k^2 = n_2^2 k^2 - k_x^2 \; ; \qquad (17)$$

$$n_{3-eff}^2 k^2 = n_3^2 k^2 - k_x^2$$

Using this concept, we can write the mode equation (13) immediately. Similarly, it is possible to write the wave equation (1) in the form,

$$\left(\frac{\partial^2}{\partial x^2} + \frac{\partial^2}{\partial z^2}\right) E_z = -(k^2 n_j^2 - k_y^2)\, E_z = -\, k_{j\text{-eff}}^2\, E_z \,, \tag{18}$$

where j=1,4 and 5 denote regions 1, 4 and 5, respectively. The effective refractive indices for these regions are now:

$$n_{1\text{-eff}}^2 k^2 = n_1^2 k^2 - k_y^2 \,;$$

$$n_{4\text{-eff}}^2 k^2 = n_4^2 k^2 - k_y^2 \,; \tag{19}$$

$$n_{5\text{-eff}}^2 k^2 = n_5^2 k^2 - k_y^2$$

Using the wave equation (18), the channel waveguide becomes a slab waveguide composed of regions 1, 4 and 5, and we may write immediately the other mode equation (15) which is identical to (9) except that the effective indices given in (19) have been used to replace the true refractive indices of the media.

The effective index method described above simplifies the problem provides a clearer picture of the wave, and more importantly, gives a solution which is as accurate as the result of an elaborate analysis. The method is derived from the fact that waves in a channel waveguide are plane waves and that k_x must be identical in regions 1, 2 and 3 and k_y must be identical in regions 1, 4 and 5. In addition, a TE wave in regions 1, 2 and 3 and a TM wave in regions 1, 4 and 5 share the same Pyonting vector $E_{x1} x H_{y1} \hat{z}$ which is the true power flow in the waveguide.

2.7 Diffused and Curved Waveguides

In a diffused slab waveguides (Figure 3a), the refractive index decreases exponentially from the top surface of the substrate, $y'=0$, into the substrate toward $y'=\infty$, roughly according to the following function:

$$n^2 = n_0^2 + \Delta\varepsilon \exp(-y'/d) \tag{20}$$

Here, we consider a Z' -system and the light wave propagates in the waveguide along z'-axis in the form of $\exp(i\omega t - i\beta z')$. With these in mind, the wave equation (1) now becomes

$$\left[\frac{\partial^2}{\partial y'^2} + k^2 \{n_0^2 + \Delta\varepsilon \exp(-y'/d)\} - \beta^2\right] H_{z'} = 0 \tag{21}$$

A similar equation would be written for $E_{z'}$, if we were considering a TM wave. We expect that the fields will concentrate near the surface of the substrate, where the refractive index is enhanced the most by diffusion. The above equation has been solved by Tien using the WKB approximation and by Conwell using Bessel functions. These theories are given as reading materials. The possibility of using Bessel functions as a solution suggests that the wave equation in cylindrical coordinates can be transformed in a form similar to (21). We will show this when we consider a disc waveguide below.

The simplest curved waveguide is a thin disc of refractive index n and of radius R, formed on a substrate of index n_0 (Figure 3b). It is tutorial to study this simple waveguide although the same solutions apply to all curved waveguides. The vector wave equation in cylindrical coordinates is rather complicated. Fortunately, it is possible to separate E_z or H_z from the rest of the field components in the equation. We have therefore,

$$\nabla^2 E_z = -\omega^2 \mu\varepsilon n^2 E_z = -k^2 n^2 E_z;$$

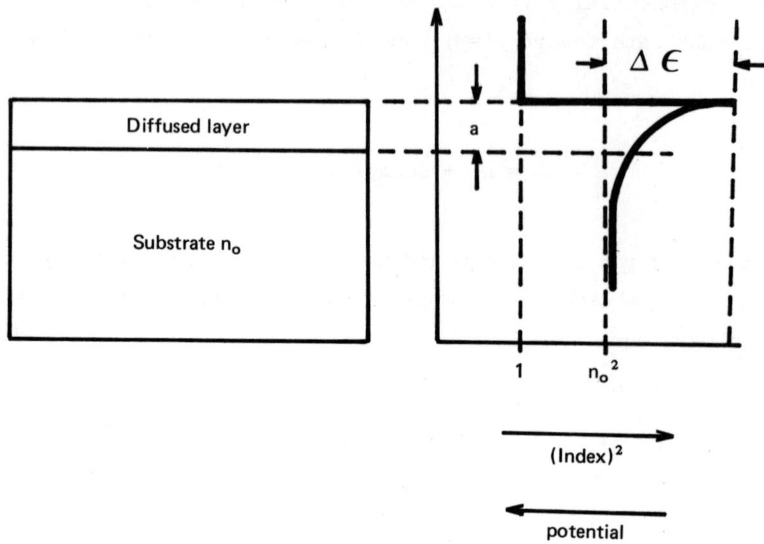

Figure 3a. Diffused Waveguide

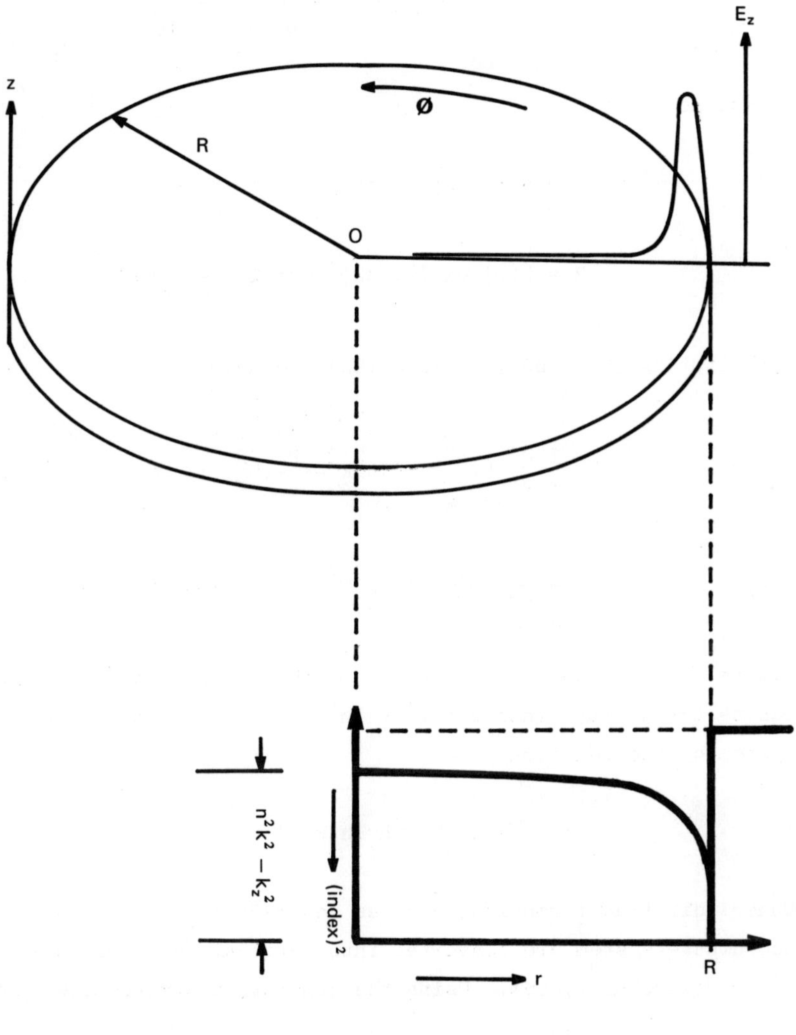

Figure 3b. Disk Waveguide

$$\nabla^2 H_z = - \omega^2 \mu \varepsilon n^2 H_z = -k^2 n^2 H_z , \tag{22}$$

where

$$\nabla^2 = \frac{1}{r}\frac{\partial}{\partial r}\left(r\frac{\partial}{\partial r}\right) + \frac{1}{r^2}\frac{\partial^2}{\partial \phi^2} + \frac{\partial^2}{\partial z^2} \tag{23}$$

We assume

$$\begin{aligned} E &= A(r)\exp(-i\ \beta R\phi)\exp(i\ \omega t - ik_z z) ; \\ H &= B(r)\exp(-i\ \beta R\phi)\exp(i\ \omega t - ik_z z) \end{aligned} \tag{24}$$

Substituting (23) and (24) into (22), we have

$$\begin{aligned} &\frac{1}{r}\frac{\partial}{\partial r}\left(r\frac{\partial E_z}{\partial r}\right) + [n^2k^2 - k_z^2 - (\beta\frac{R}{r})^2]E_z = 0 ; \\ &\frac{1}{r}\frac{\partial}{\partial r}\left(r\frac{\partial H_z}{\partial r}\right) + [n^2k^2 - k_z^2 - (\beta\frac{R}{r})^2]H_z = 0 \end{aligned} \tag{25}$$

Figure 4 shows that $W = r\exp(i\phi)$ in the cylindrical coordinate system can be transformed into $U = y' + iz'$ of the rectangular coordinate system by the relation

$$U = R\ln(W/R) \tag{26}$$

Under this transformation, circles and radial lines in the cylindrical coordinate system are converted into vertical and horizontal lines in the rectangular system. Using the same transformation, we put

$$y' = R\ln(r/R)$$

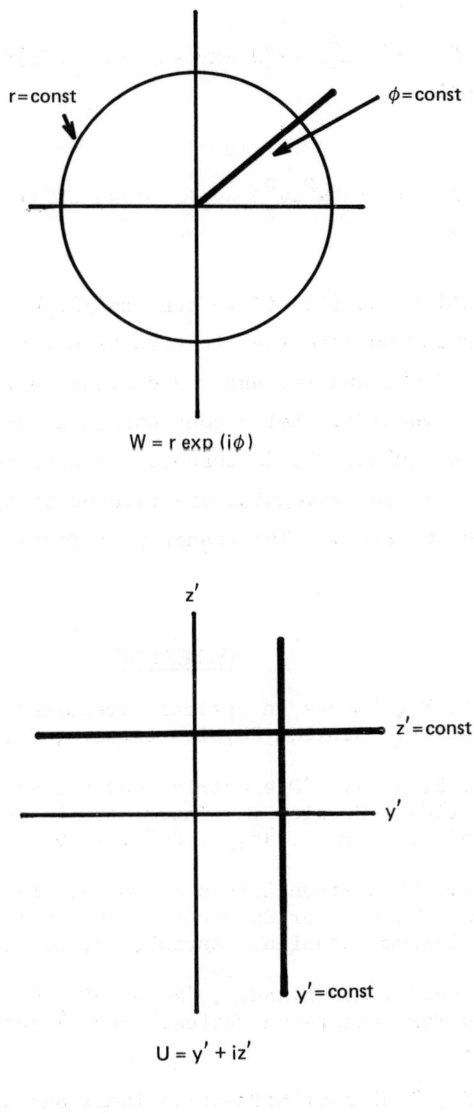

Figure 4. Conformal transformation
$U = \ln(W/R)$

in equation (25) and obtain

$$[\frac{\partial^2}{\partial y'^2} + \{(n^2k^2-k_{z'}^2)\ \exp(-2y'/R)\ -\beta^2\}]E_{z'}=0$$

$$\frac{\partial^2}{\partial y'^2} + \{(n^2k^2-k_{z'}^2)\ \exp(-2y'/R)\ -\beta^2\}]H_{z'}=0 \qquad (27)$$

They are identical to (21) if we take $d=R/2$, $k^2\Delta\varepsilon=(n^2k^2-k_{z'}^2)$ and $n_0=0$. It is interesting to note that in this transformation r and ϕ correspond to y' and z', and r=0 and r=R correspond to $y'=-\infty$ and y'=0. In the diffused waveguide, fields concentrate at the surface y'=0. Here in the disc waveguide, fields concentrate near the edge of the disc. The modes of the disc waveguide are related to the "whispering-gallery" modes of Lord Rayleigh. The reader is referred to Reference #8 for details.

REFERENCES

5. Conwell, E. M., "Modes in Optical Waveguides Formed by Diffusion," Applied Physics Letters, 23, p. 328, 1973.

6. Marcatili, E. A. J., "Dielectric Rectangular Waveguide and Directional Coupler for Integrated Optics," Bell System Technical Journal, 48, p. 2071, 1969.

7. Marcuse, D., "Radiation Losses of Dielectric Waveguides in Terms of the Power Spectrum of the Wall Distortion Function," Bell System Technical Journal, 48, p. 3233, 1969.

8. Sheem, S. and J. R. Whinnery, "Modes of a Curved Surface Waveguide for Integrated Optics," Wave Electronics, 1, p. 105, 1974.

9. Tien, P. K., "Rules of Refractive Index and a Potenital Well Model of the Optical Waveguides," Radio Science, 16, p. 437, 1981.

3. THIN FILM PRISMS AND LENSES, METHOD OF LIGHT GUIDING INTERCONNECTIONS AND INTEGRATED OPTICAL CIRCUITS FOR SIGNAL PROCESSING

3.1 Introduction

While we are discussing modern optical science, we should not forget great works in the past, particularly, the Fermat principle and the Huygen principle which describe so beautifully the nature of light wave propagation. There are many ways to describe the Fermat principle, one of them is:

"The paths taken by light rays in going from one point to another through any set of media are such as to render all optical paths equal".

The optical path is an integral of n ds, where n is refractive index of the medium in which an optical wave traces a differential path of ds. The Huygen principle is simpler, it states:

"Each point on a wave front may be regarded as a new point source of waves".

The Snell law of refraction, the related phenomenon of total internal reflection, the lens law for focusing, and in fact the entire geometric optics may be derived from these principles. Of course, the two-dimensional optics which we shall discuss is another extension of these fundamental principles.

Speaking of the two-dimensional optics, we return to the wave equation (1)

$$\left(\frac{\partial^2}{\partial x^2} + \frac{\partial^2}{\partial y^2} + \frac{\partial^2}{\partial z^2}\right) E = -k^2 n^2 E$$

Replacing $(\partial^2/\partial z^2)$ by $-k_z^2$, we have

$$\left(\frac{\partial^2}{\partial x^2} + \frac{\partial^2}{\partial y^2}\right) E = -(k^2 n^2 - k_z^2)\, E = -k^2 n_{eff} E \qquad (30)$$

We may write the above equation as long as k_z is uniform over the entire x-y plane. Here n_{eff} is the effective refractive index of the two-dimensional wave equation (30). If the medium considered is a slab waveguide, $\hat{z}$ is along the thickness of the film and z-y plane is the surface of the waveguide. In that case, k_z depends upon the thickness of the film and the refractive indices of the film and substrate. We can ignore all the detailed parameters of the waveguide. So far the wave motion in (30) is considered, a single quantity n_{eff} completely characterizes the medium, and thus the wave propagation.

In fact, it is not necessary for the light wave to propagate in the x-direction as we have assumed in LECTURE 1, the wave can propagate in any direction along the x-y surface of the waveguide. We thus consider the slab waveguide as though it were a two-dimensional free space of refractive index, n_{eff}. Naturally, all the laws of wave propagation used in 3-dimensional space will also apply to the two-dimensional space.

When two slab waveguides join together at a boundary a-a', the effective indices, n_{eff-A} and n_{eff-B}, may be quite different, because either different thicknesses or different refractive indices are used in the two waveguides. In that case, refraction and total reflection will occur according to the Snell law as a light wave crosses the boundary a-a'. These phenomena are used to construct thin-film prisms and lenses.

It was learned early in 1971 that all deposited films have tapered edges. Some of the tapers are quite long covering many optical wavelengths. When the tapered edges of two light-guiding films superpose one another, they form a smooth junction and a light wave can freely propagate through the junction without scattering. These tapered edges of the films resemble to the tapered waveguides in microwave circuits where, they are frequently used for matching impedances of microwave devices.

We also learned quite suprisingly that, not only the wave propagates freely through a tapered junction, but also it maintains a same mode order before and after the transition. We call it the "adiabtic" nature of wave propagation in the sense that, the wave energy in a waveguide mode is conserved throughout an optical circuit provided that all the devices in the circuit are interconnected by tapered junctions. This is important in optical communication, since each waveguide mode carries a channel of information. When wave energy in each mode is conserved, cross-talk between channels will not occur. It is also important so far the construction of the circuit is considered, since deposited films have tapered edges which form tapered junctions quite easily.

In the last few years, integrated optical circuits have been investigated for signal processing. For example: a 4-bit 275 MS/s electrooptic A/D converter was reported by Lincoln Laboratory and a time-integrating correlator using Bragg diffraction by Professor Tsai and his students. Although the technology presently available in integrated optics has not reached the perfection needed for constructing complex circuits. The optical circuits will eventually win because of enormous bandwidths they have.

3.2 Optical Circuits in a Multimode Fiber System

Because of the adiabtic nature of light wave propagation in optical circuits discussed in the last section, it is difficult to couple light from a multi-mode fiber to a channel or slab waveguide of the optical circuit. All thin-film couplers work best with the fundamental mode of the fiber which has a simple Gaussian field distribution, leaving other modes more or less uncoupled. It is, however, possible to multiplex light from several single-mode fibers into a slab or channel waveguide. One way to do this is to use a prism coupler. By placing several single-mode fibers at the synchronous angles of the coupler, light from the fibers can be coupled into different waveguide modes of an optical circuit. We may also reverse the process, dispensing light of different waveguide-modes to single-mode optical fibers.

We have mentioned earlier in LECTURE 1, in order to work with a multimode fiber system presently used for communication, detectors must be placed at the front end of an optoelectronic circuit. Light from the multimode fiber is first converted to electricity and signal is processed at the electrical level. Such circuits are called PINFET since they involve a p-i-n diode detector coupled to field-effect-transistor amplifiers.

REFERENCES

10. Reinhart, F. K. and R. A. Logan, "GaAs-AlGaAs Double-Heterostructure Laser with Taper Coupled Passive Waveguides," Applied Physics Letters, 26, p. 516, 1975.

11. Technical Digest of 1982 Topical Meeting on Integrated and Guided Wave Optics, Pacific Grove, California (Published by The Optical Society of America).

12. Tien, P. K. and R. J. Martin, "Experiments in Light Waves in a Thin Tapered Film and a New Light-Wave Coupler," Applied Physics Letters, 18, p. 398, 1971.

13. Tien, P. K., R. J. Martin and G. Smolinsky, "Formation of Light-Guiding Interconnections in an Integrated Optical Circuit by Composite Tapered-Film Coupling," Applied Optics, 12, p. 1909, 1973.

14. Tien, P. K., S. Riva-Sanseverino, R. J. Martin and G. Smolinsky, "Two-layered Construction of Integrated Optical Circuits and Formation of Thin-Film Prisms, Lenses and Reflectors," Applied Physics Letters, 24, p. 547, 1974.

4. COUPLING OF TWO WAVEGUIDES AND ITS APPLICATIONS AS MODULATORS AND SWITCHES

4.1 Introduction

When two light waves are coupled, interference phenomena will occur. This happens when two identical channel waveguides are placed side by side so that the fields in one waveguide penetrate into the other waveguide. Such an arrangement is called a directional coupler. Two light waves propagating as a TE and a TM wave respectively in a waveguide may also be coupled to one another if the waveguide is made of an anisotropic, or a magnetooptic or an electrooptic material. An example of this is the magnetooptic polarization converter.

Three parameters quantitiatively characterize the coupling phenomena; they are:

L = the length of the region where the two waves are coupled;

κ = the coupling coefficient which specifies coupling strength;

$\Delta\beta$ ($=2\delta$) = the difference in phase constants of the two waves.

Consider a directional coupler, L is the length of the region where the two waveguides are located side by side. The spacing between two waveguides determines the coupling coefficient, K, which in fact decreases exponentially with the spacing. Consider waves in the waveguides which propagate as $\exp(i\omega t - i\beta_1 z)$ and $\exp(i\omega t - i\beta_2 z)$, respectively, $\Delta\beta=2\delta=\beta_1-\beta_2$. There are two types of coupling. We call it "codirectional" when two waves propagate in the same direction, and "contradirectional" when they propagate in opposite directions. In the former case the amplitudes of the two waves vary sinusoidally with distance and in the latter case, they vary exponentially with distance.

Here, we shall only outline briefly what we are going to

discuss in this lecture. We start by going back to the wave equation (1) and show that this second order differential equation can be written into two first order differential equations, one for the wave propagating in the forward (+z) direction and the other for the wave propagating in the backward (-z) direction. We then derive coupled-mode equations for the codirectional and contra-directional couplings. In the case of codirectional coupling, the discussion includes an magnetooptic modulator and use of serpentine circuit for phase matching. We will also investigate an electrooptic directional coupler switch in the COBRA configuration and a stepped $\Delta\beta$ reversal arrangement. For the case of contra-directional coupling, we will study a Bragg reflector and an acoustooptic lightwave modulator.

The coupled mode method used to solve all these problems is a first order perturbation theory. Mathmatical solutions of the theory should be straight-forward. However, because of the involvement of phases of the waves, there are times that the results are confusing. For that reason, an interesting graphic method will be used to explain simple as well as complicated cases.

4.2 Graphic Method

The graphic method I propose may be illustrated by considering the codirectional case for which, the coupled mode equations are

$$R' - i\,\delta R = -i\kappa S \;;$$

$$S' + i\,\delta S = -i\kappa R \tag{31}$$

Here, the prime indicates the differentiation with respect to the distance z, and R and S are the amplitudes of two coupled waves. δ and κ have been defined before. Mathmatical solutions of (31)

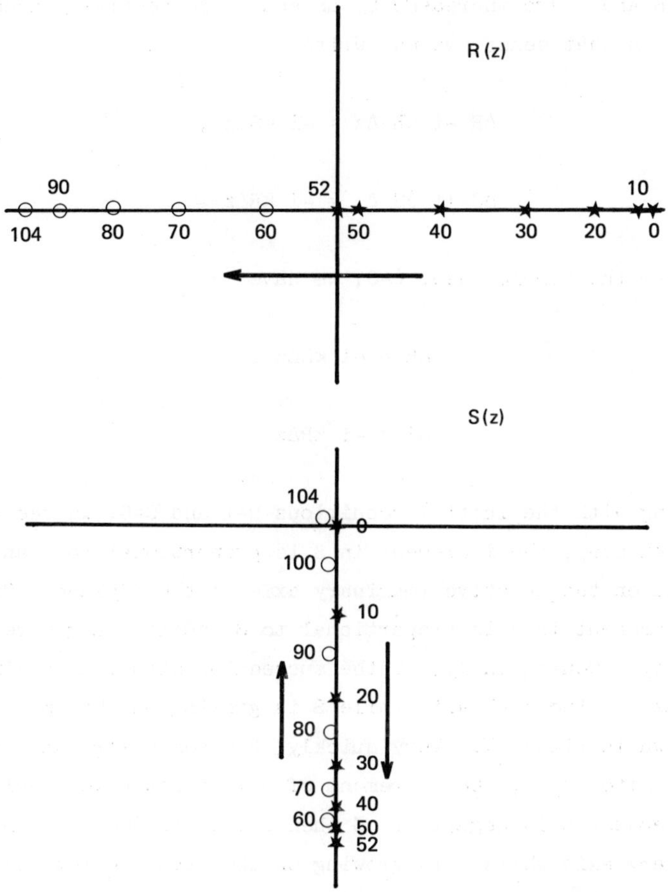

Figure 5. Graphic solution for codirectional case
$\kappa = 0.03, \ \delta = 0.00$

are given in the Kogelnik's paper (see Reference 17) We draw in Figure 5 two complex planes, one for R and one for S. Let us divide the distance z into a number of equal steps of Δz. In each step, R and S are increased by ΔS and ΔR respectively according to (31). In that sense, we may write

$$\Delta R - i\ \delta R\ \Delta z = -i\ \kappa S \Delta z\ ;$$

$$\Delta S + i\ \delta S\ \Delta z = -i\ \kappa R \Delta z \qquad (32)$$

Consider the simple case, $\delta=0$, we have

$$\Delta R = -i\ \kappa S \Delta z\ ;$$

$$\Delta S = -i\ \kappa R \Delta z \qquad (33)$$

Starting with the initial conditions R=1 and S=0, we see from (33), for each step, the increment in S is proportional to R and may be plotted on the negative imaginary axis of the S plane. Similarly, the increment in R is proportional to S and is a negative real quantity. Consequently, in the succeeding steps, R continues to decrease on the real axis while S is growing on the imaginary axis as shown in Figure 5. Very quickly, R becomes zero and then negative. After that, the increment of S is a positive imaginary quantity because R is negative. S then starts to decrease on the imaginary axis while R is growing on the negative real axis. We thus demonstrate that R is a cosine function in z and S is a sine function in z. We can draw similar diagrams for $\delta \neq 0$. During the lecture, we will also work out the graphic solution for the more complicated stepped $\Delta\beta$ reversal switch. The graphic method differentiates interference caused by coupling from that due to the phases of the waves and thus provides a clearer picture of the coupling process. The reader is encouraged to use this method to invent new schemes of coupling.

REFERENCES

15. Alferness, R. C., "Guided Waves Devices for Optical Communication," IEEE Journal of Quantum Electronics, QE-17, p. 946, 1981.

16. CRC Handbook of Lasers with Selected Data on Optical Technology, Published by the Chemical Rubber Company, 1971.

17. Kogelnik, H., "Coupled-Wave Devices" in "Fiber and Integrated Optics" Edited by D. B. Ostrosky, Plenum Publishing Company, 1979.

18. Kogelnik, H. and R. V. Schmidt, "Switched Directional Couplers with Alternating $\Delta\beta$," IEEE Journal of Quantum Electronics, QE-12, p. 396, 1976.

19. Tien, P. K. and D. P. Schinke and S. L. Blank, "Magneto-optics and Motion of the Magnetization in a Film-Waveguide Optical Switch," Journal of Applied Physics, 45, p. 3095, 1974.

20. Yaamamoto, S., Y. Koyamada and T. Makimoto, "Normal Mode Analysis of Journal of Applied Physics," 43, p. 5090, 1972.

21. Yariv, A., "Coupled Mode Theory for Guided-Wave Optics," Journal of Quantum Electronics, QE-9, p. 919, 1973.

5. INTEGRATION OF OPTICS WITH ELECTRONICS

5.1 Introduction

We should realize that most of optical devices discussed earlier such as lasers, detectors, modulators and switches, are optoelectronic devices. They will not function without electricity either in the form of d-c bias or a-c switching signal, or both. It is, therefore, highly desirable to integrate optical devices and electronic circuits on a same substrate. There are definite advantages for such integration: First of all, numerous inter-connects between electronic and optical components are needed for feedback and stabilization, for d-c bias and regulation,for error detection and pulse shaping, and for modulation and switching. Integration would reduce parasitic reactances, ease the fabrication and thus, increase the reliability and reduce the cost. More importantly, it provides more freedom in circuit design. As we have mentioned in Lecture I, an optical waveguide has an infinite bandwidth as compared with the RC constant of the electronic circuit. It is economical to operate the signal at the optical level when a large bandwidth is involved. The traditional concept of circuit design is changed when the circuit contains both optical and electronic devices. Finally, we should consider optoelectronic integration is a first step toward the modular construction of an optical communication system which should be easy to install and to repair and free from environmental interferences. It is true that our long term goal is to provide a large-scale integration of optoelectronic circuits for new services and greatly expanded bandwidth requirements envisioned for the "information age". At the present time, however, integrating even a relatively small numbers of photonic devices, transistor and functions into small chips should benefit present-day light-wave systems both technically and economically. Because of this and inspired by the scientific interest in general, research on optoelectronics is greatly

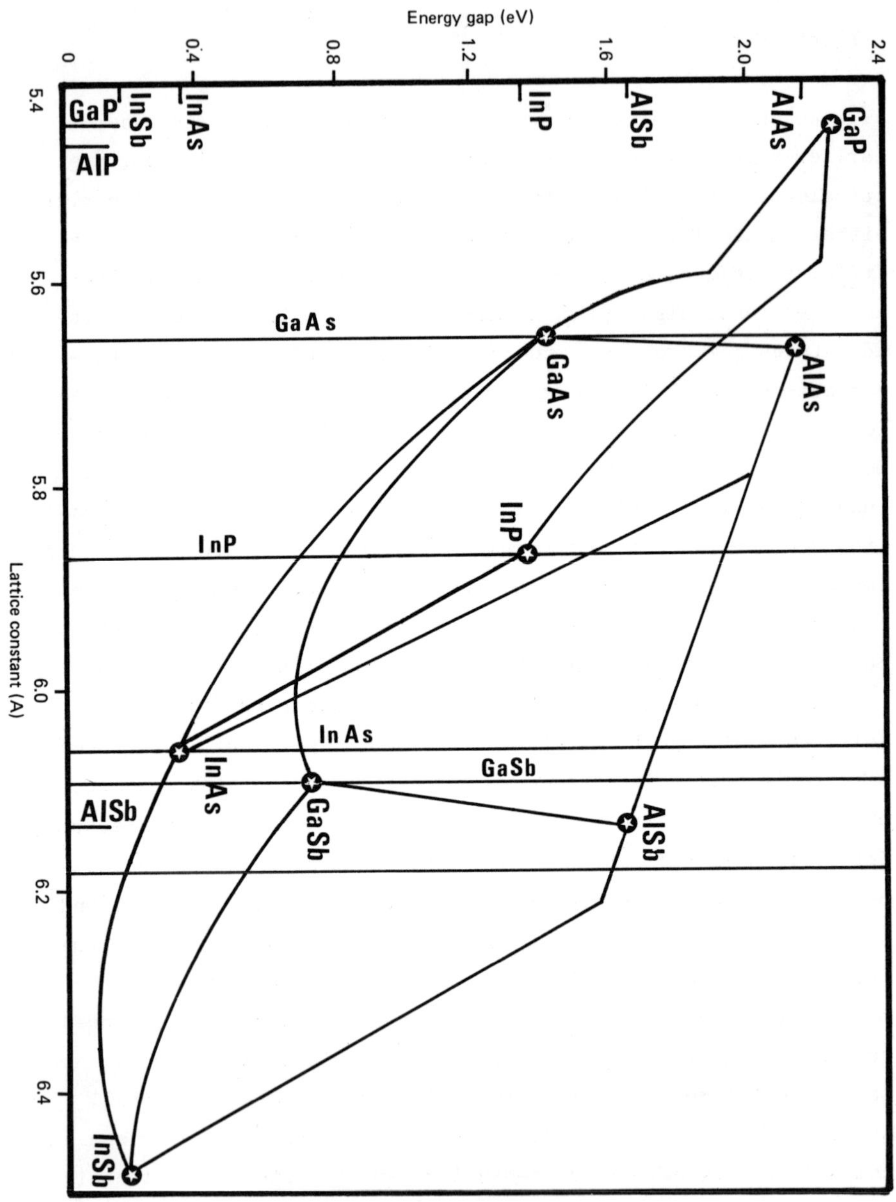

Figure 6. Materials for III-V alloy compounds

accelerated.

Therefore, we are searching for materials and processing technology which can be used for both optics and electronics. Various material systems we shall consider are III-V alloy compounds shown in Figure 6. Use of these materials for lasers and detectors have already been discussed in the lectures given by Dr. M. Nakamura. The figure shows the energy of band gap in electron volts (left vertical scale) versus lattice constant in angstroms (horizontal scale) of all III-V compounds and their alloys. When lasers are made from these materials, the wavelength of the radiation in microns is determined from the bandgap energy as it is indicated in the right vertical scale. In general, the material is transparent to a radiation, if it has a wavelength longer than the wavelength indicated by the bandgap. The same material will then absorb radiations which have shorter wavelengths. For that reason, we can choose one III-V compound for the detector which absorbs a radiation and another compound for the waveguide which is transparent to the same radiation. In this figure, the binaries such as GaAs, AlAs, GaP, InP, etc., are marked in bold letters. The tie lines between the binaries are the ternaries. For example, the line connecting GaAs to AlAs shows the bandgap energy versus lattice constant of the ternary, $Al_xGa_{1-x}As$. For x=0, it is the binary, GaAs. The vaule of x increases from 0 to 1 as we move along the line from GaAs to AlAs. The first portion of the tie line is a solid curve indicating that the material has a direct energy gap. The line becomes dashed as the bandgap turns to be indirect. Lasers must be constructed from direct bandgap materials whereas indirect bandgap materials may be used for waveguides and detectors.

Any material inside a polygon, for example, the polygon GaP-GaAs-InAs--InP, is a quarternary. In this case, the quarternary is $Ga_xIn_{1-x}As_yP_{1-y}$, where both x and y vary from 0 to 1. Other quarternaries may also be identified by the polygons which surround them. These polygons are shown in Figure 6. For material systems based on GaAs and InP, intensive research has been carried out for lasers and detectors. All the semiconductor lasers and detectors are made of multi-layer structures. Each layer is a different alloy compound. The technology for growing different compounds one on top of the other is called epitaxy. To insure a good crystalline quality, the materials used in a multilayer structure must have same lattice constant and must thus lie on a same vertical line in Figure 6. For example, all $Al_xGa_{1-x}As$ ternaries have nearly the same lattice constant as GaAs and thus may be grown epitaxially on GaAs substrates. The technology involved is called GaAs technology. Similarly, the technologies involved in growing lattice-matched $Ga_xIn_{1-x}As_yP_{1-y}$ on InP and $Al_xGa_{1-x}As_ySb_{1-y}$ on GaSb are called InP and GaSb technologies, respectively. We have therefore, several material systems to choose from.

5.2 GaAs - Based Optoelectronics

As shown in Figure 6, GaAs/AlGaAs lasers cover the wavelength from 0.64 to 0.86 μm. These lasers have been used in all the early optical communication systems. Use of GaAs to form lasers and detectors will be discussed in the Nakamura's lecture. Here we will concentrate our effort on the formation of transistors and electronic circuits side by side with the optical devices. Historically, GaAs technology is known for ultrahigh speed electronics. In order to provide a broad background to our later discussion, we will outline below material properties, and the advantages and disadvantages of GaAs technology particularly in comparison with the well established silicon electronics.

Electron mobilities in GaAs are about 8500 cm^2/Vs in undoped materials and 4000-5000 cm^2/Vs in typical MESFET n-channels ($N_d=10^7$ cm^{-3}) at room temperature. These GaAs electron mobilities are about 6 times higher than those for correspondingly doped bulk silicon, and nearly an order of magnitude larger than typical silicon MOS n-channel mobilities under strong inversion conditions. Even higher mobilities have been obtained in GaAs at lower temperature and in modulation-doped materials. Because of this high mobility and the 0.5 -1.0 μm gate lengths used in microwave FETs, GaAs analog amplifiers have been used extensively in the 2-18 GHz range. In digital circuits, propagation delays as low as 30 ps/gate at room temperature and 17.5 ps/gate at 77°K have been achieved. In general, one can gain a factor of 2- 10 better in speed in GaAs than in silicon.

Another advantages of GaAs is the availability of the high resistivity material (10^{8} ohm-cm) when it is compensated with Cr or O_2, or other deep level acceptors. Active devices can be formed by direct ion-implantation into the high resistive material. This not only eases the problem of providing isolation between transistors, but also eliminates capacitances between substrate and source and drain. The semi-insulating substrate is also helpful in reducing interconnect capacitances in an integrated circuit. We have mentioned in Lecture I that the RC constant of the interconnects is significant in a high speed circuit.

Lack of reliable oxides on GaAs, the Schottky-barrier field effect transistors (MESFET) are usually used in GaAs circuits. While these devices are excellent depletion mode (normally-on) transistors, tight control in logic voltage swing is required when they are used as enhancement mode (normally-off) FETs. In the case of enhancement mode operation, the built-in potential of the Schottky metal gate prevents current flow at V_{gs}=0. However, it draws excessive current when the device is forward biased above 0.6 to 0.8 volts. The logic voltage swing is thus limited to 0 to 0.5 volts. Such a small voltage swing requires very tight controls in circuit fabrication, although, circuits containing over 6000 active devices have been constructed.

It is therefore not surprising that MESFETs have been used by Caltech's group for integration with lasers. The work of Caltech's group on the GaAs-based optoelectronics has been summarized very well by Bar-Chaim, et. al., which is included in the "Reading Material" as a part of these notes. In addition, a large number of publications are available on GaAs-based monolithic integrated optics. A summary of these publications is also included in the "Reading Material". With these materials as background, we may now proceed to discuss InP-based optoelectronics.

5.3 Long Wavelength Communications and InP Technology

Optical fibers used for light wave communication have a lower loss and zero dispersion near the wavelength of 1.3 μm (see Kao's lecture) which is much longer than the wavelength, 0.8-0.85 μm, used in earlier systems. For that reason, light wave communication systems operated at wavelengths between 1.0 and 1.6 μm are called long wavelength communication. We have seen in Figure 6 that lasers made from quarternaries, $Ga_xIn_{1-x}As_yP_{1-y}$, cover the above wavelength range, and these quaternaries are lattice-matched to InP substrates if $y/x \cong 2.2$. Because of this, a great deal of research has been focused on InP-based technology. The InP technology is still in an elementary stage. In many ways regarding diffusion, ion-implantation, Schottky barrier and metal-to-semiconductor contacts, it is very different from GaAs technology. Inspite of this, long-life $Ga_xIn_{1-x}As_yP_{1-y}$ lasers have been demonstrated both in United States and Japan in the 1.2 -1.55 μm wavelength range, and InGaAs detectors which have a dark current, 1 nA, and capacitance, 1 pF, competes fairly in performance with Si-avalanche photodetectors used in earlier sytems.

In an extensive study of alloy scattering, it was first proposed by Leheny et. al. that $In_{.53}Ga_{.47}As$ which is lattice-matched to InP has the largest electron mobility in the entire lattice-matched $Ga_xIn_{1-x}As_yP_{1-y}$ system (Figure 7). In this system, the room temperature electron mobility varies from 3800 cm^2/Vs for InP up to 11,000 cm^2/Vs for $In_{.53}Ga_{.43}As$ (carrier concentration $2.4X10^{16} cm^{-3}$).

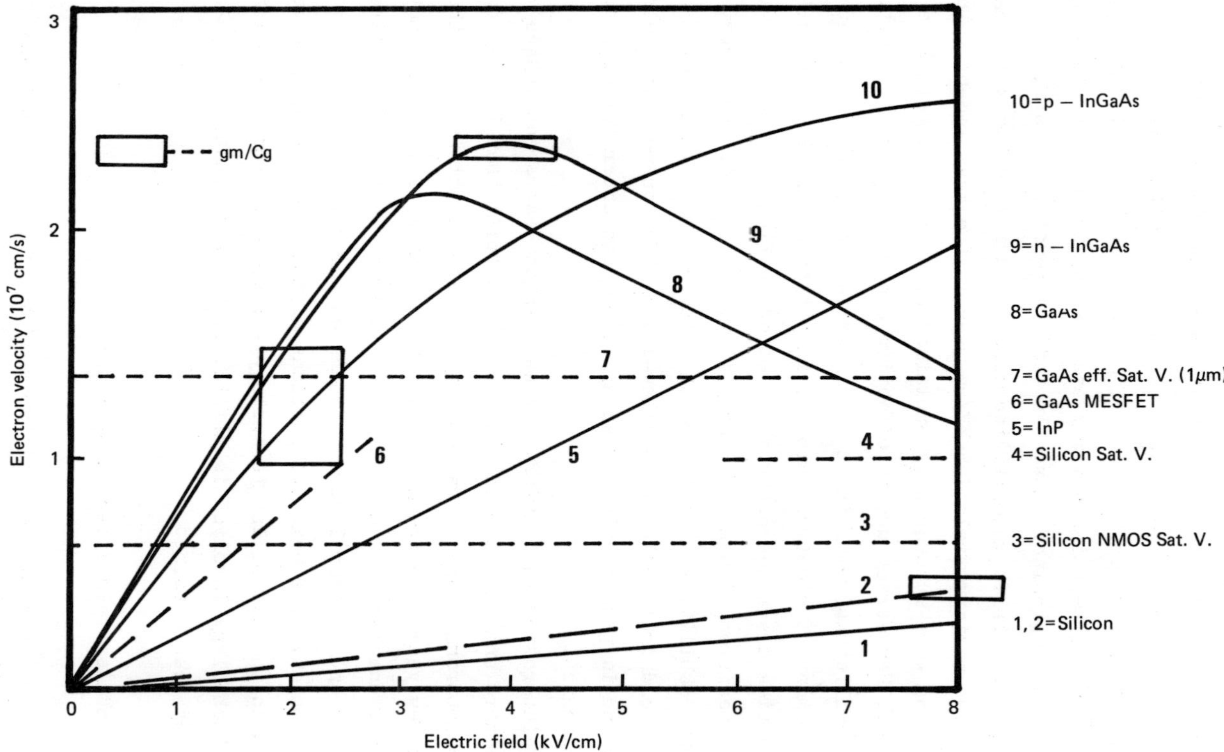

Figure 7. Electron velocity vs field

The high-field electron drift-velocity in InGaAs has also been measured. It is interesting that electrons behave quite differently in n and p types of materials. In the n-type InGaAs, the electron drift velocity reaches a peak of 2.5×10^7 cm/s at 3.75 kV/cm, exhibiting current instability associated with negative differential resistance typical of GaAs. In the p-type of materials, however, the electron drift velocity increases steadily to 2.6×10^7 cm/s up to the maximum field 7.5 kV/cm used in the measurement. These phenomena were subsequently explained by a group at Bell Labs in Holmdel, New Jersey due to effective energy exchange of electron-hole collisions resulting a lower carrier temperature and thus an enhancement of electron velocity in p-type materials. The drift velocities discussed above are plotted in Figure 6 together with those measured in GaAs, InP and Si. The slopes of these curves are the mobilities. We notice that InGaAs has both larger high-field drift-velocity and low-field electron-mobility than GaAs and is thus considered as a serious contender for high speed electronics.

Interested in GaInAsP long wavelength lasers and InGaAs detectors and electtronics, it is only natural that InP-based optoelectronics has attracted a great deal of attention in the last few years. Since both GaInAsP and InGaAs are lattice matched to InP, one can grow epitaxially a layer of GaInAsP for optical devices and another layer of InGaAs on the same substrate for electronics. The next two sections of this lecture will be devoted to the use of InP technology for optoelectronic integration. The technology involved is difficult, but excellent results have already been obtained.

5.4 InGaAs Field Effect Transistors

Figure 8 shows several types of the field-effect transistors commonly used in integrated circuits. On the top-left is the insulated gate FET. Both Si_3N_4 and a native oxide have been used

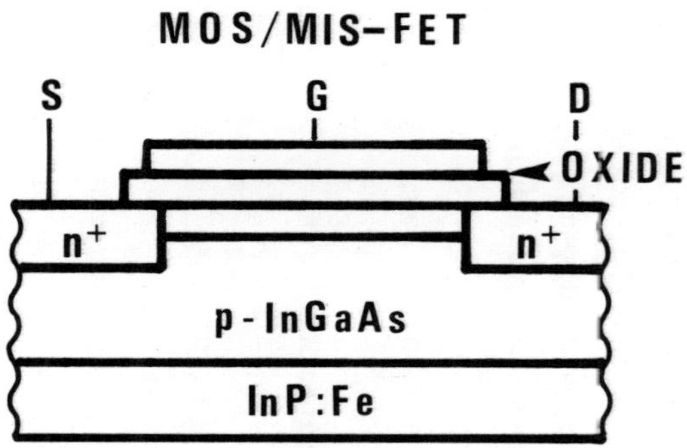

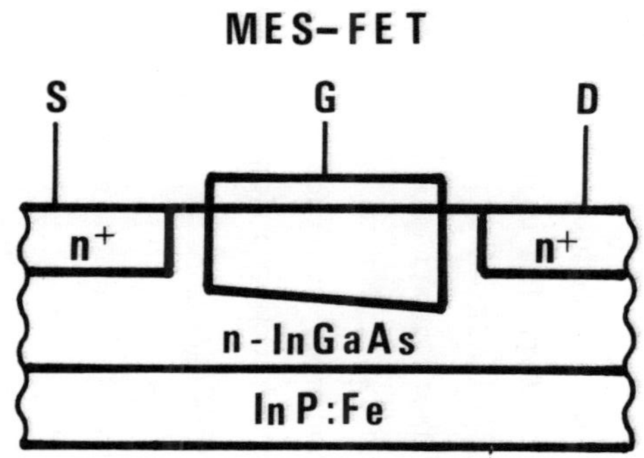

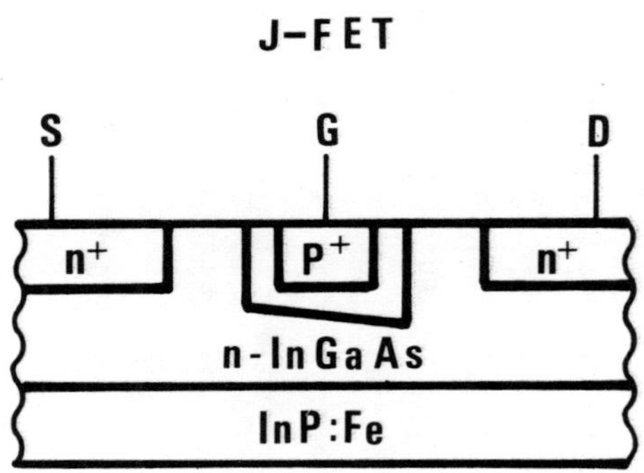

FET	Si	GaAs	InGaAs
MOS	✓		✓
MIS	✓		✓
MES	✓	✓	
J	✓	✓	✓

Figure 8. Several types of FET used in IC'S

as the gate insulators for InGaAs FETs. Si_3N_4 reported by Liao, et. al., is formed by a low temperature RF plasma enhanced CVD process with 3% silane (in Ar) and nitrogen source gases. The sample is maintained at 350°C. The resulting film has a refractive index close to 2, a breakdown field intensity of 5×10^6 V/cm and a resistivity larger than 10^{14} ohm-cm. The plasma-grown native oxide discovered by Tell, et. al., is water insoluble and can be formed on both n-type and p-type InGaAs epi-layers. The oxide resistivity is of order 10^{12} ohm-cm and breakdown field, 10^6 V/cm.

Using either of the above insulators, n-channel and also p-channel inversion mode FETs have been reported by Liao, et. al. Source and drain can be formed by diffusion, Schottky contact and recently, by ion-implantation. The threshold voltage can also be adjusted by ion-implanting the channel. In general, the fabrication process is not very different from that used for silicon ICs, although each material system has its own properties and requires special consideration. Recently, a transconductance of 40 mS/mm has been obtained by Liao. which is significantly better than that reported earlier.

5.5 Simple Integrated Optoelectronic Circuit-PINFET

Figure 9 is a PIN/JFET photoreceiver reported by Leheny, et. al., at the 1982 Topical Meeting on Integrated Optics and Guided Waves. The circuit consists of a photodiode as a detector at the left and a JFET as an amplifier at the right. A layer of unintentionally doped InGaAs was first grown on a semi-insulating InP:Fe substrate. The epi-layer is n-type and has a concentration of $0.5\text{-}4 \times 10^{16}$ cm^{-3} with an electron mobility on the order of 9000 cm^2/Vs. The p-layer of the PIN photodiode shown at the left and the gate of the JFET shown as a narrow line, were formed with a single diffusion of Zn through a Si_3N_4 mask at 550°C for 15 minutes. Typical junction depth was 1.0-1.3 μm leaving 0.7-1.0 μm for the FET channel thickness. The photodiode has an area of 150x100 μm, and the gate is

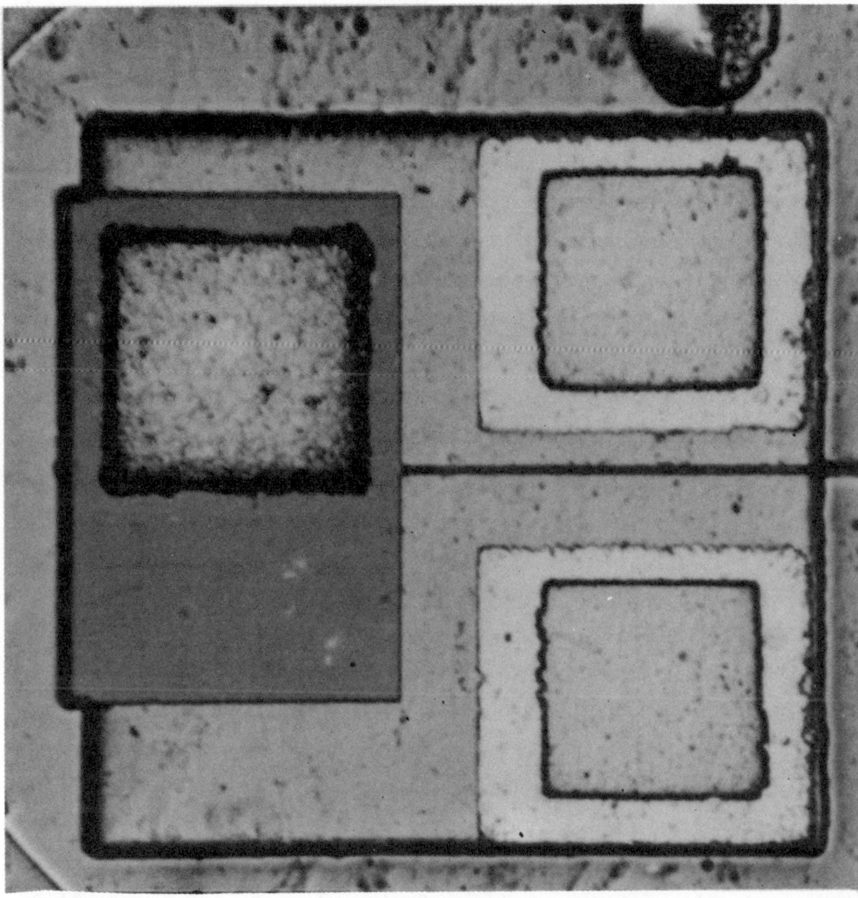

Figure 9. PIN/JFET photoreceiver

5-7 μm long and 120 μm wide. After removing Si_3N_4, electric contacts were applied to source and drain and to the photodiode by plating Au-Ni-Sn for the n-contact and Au-Zn for the p-contact followed by sintering at 410^oC.

The photodiode in the above PINFET has a dark current less than 40 nA for bias up to -3 Volts. The n-layer under the gate is fully depleted at a reverse bias of 12 volts. The transconductance of the FET is 58 mS/mm. In a separate experiment, a transconductance as large as 125 mS/mm has been obtained.

It is obvious that we face a number of difficulties in constructing optoelectronic circuits. Consider this PINFET circuit as an example, it is difficult to control diffusion in InGaAs. If the speed of the circuit is our primary concern, diffusion cannot be used in forming short gates. The trend is to use an ion-implant for the photodiode and gate junction and another ion-implant for the source and drain. Very little is known about ion-implant in InGaAs. Particularly, InP will decompose at a temperature above 600^oC. Post annealing after ion-implant may be a problem. If more transistors are to be integrated into this circuit, schemes for isolation between transistors have to be provided. Extensive study on metalization and field oxides should also be carried out. Obviously, we have more questions to be answered than we have answers ready for them. Much more work has to be done before a practical optoelectronic circuit can be realized.

In silicon circuits, logic signals are carried by the diffusion and polysilicon lines. The crossings between the polysilicon and diffusion lines are transistors. In the PINFET circuit, we could etch conductive InGaAs epi-layer for the line to propoagate logic signal and use ion-implant line with refractive metal on the top for controlling the logic. This would greatly reduce the RC constant of the interconnects, whereas in silicon circuits, the diffusion or ion-implant line is always at the lowest level of the structure and little can be done to improve its RC constant.

The availability of the semi-insulating InP substrates in

these circuits is an asset for reducing the RC constant of the circuit. The possibility of growing multilayer structure with a different composition for each layer by molecular beam epitaxy is another asset in regard to the processing layout. We are confronted with new opportunities as well as difficulties. We are still at the starting point of the learning curve. However, the following is always true: Once we learned how to etch, to ion-implant, to diffuse impurities, to form contacts, and to deposit metal lines, gate, and field oxides, fabrication of an optoelectronic circuit should not be any more difficult than the fabrication of a silicon circuit. We are fortunate to witness the beginning of a new era of integrated circuits which are fundamentally different from those we know before.

REFERENCES

22. InGaAs/SiN N-Channel and P-Channel Inversion Mode MISFETs, 1981 IEEE IEDM Technical Digest, p. 637.

23. Nadav Bar-Chaim, Israel Ury and Amonon Yariv, "Solid State Integrated Optoelectronics," IEEE Spectrum, 19, p. 38, 1982.

24. Leheny, R. F., R. E. Nahory, J. C. DeWinter, R. J. Martin, and E. D. Beebe, "An Integrated PIN/JFET for Long Wavelength Optical Receiver," 1981, IEEE IEDM Technical Digest, p. 276.

25. Tien, P. K. and J. A. Giordmaine, "Integrated Optics: Putting It All Together," Bell Laboratories Record, February, 1981 Issue, p. 38.

QUANTUM ELECTRONICS, PICOSECOND LASER PULSES AND APPLICATIONS*

Erich P. Ippen
Department of Electrical Engineering and Computer Science
and Research Laboratory of Electronics
Massachusetts Institute of Technology
Cambridge, MA 02139, USA

ABSTRACT

A review is given of recent work in the generation of picosecond and subpicosecond light pulses. The necessary laser system and measurement techniques are described. Some applications of picosecond pulses to the characterization of ultrafast processes are discussed. Finally advances in semiconductor lasers capable of picosecond pulses are described.

TABLE OF CONTENTS

*Work supported in part by the Joint Services Electronics Program and by the National Science Foundation

1. ULTRASHORT PULSE GENERATION AND PROPAGATION

1.1 Introduction

Short flashes of light have long provided a means for capturing events in time and for studying high speed motion (Shapiro, 1977). As shorter light pulses have become available, an increasing variety of physical phenomena have become amenable to study and new areas of research have opened up. Strobe photography, pioneered (Edgerton, 1932) in the early thirties, offers microsecond temporal resolution which can be used to freeze the most rapid mechanical motion of microscopic objects. Scientific use of these high speed flash techniques pushed ahead rapidly in the fifties after their usefulness for initiating and studying fast photo-chemical reactions was demonstrated (Porter, 1950). With the invention of the laser even more rapid advance occurred. The dramatic increases in optical intensity and monochromaticity made possible by lasers were accompanied by a shortening of pulse durations from the nanosecond (10^{-9} sec) to the picosecond (10^{-12} sec) regime. Exciting opportunities for accurate study of previously unresolved, fast photoprocesses in physics, chemistry and biology were opened up. The possibility of enormously expanding the bandwidth of communication technology was also recognized.

In the decade between 1965 and 1975 major advances were made in the measurement, control, and utilization of ultrashort light pulses. With measurement required on a timescale too fast for conventional electronics, new techniques had to be invented. Interestingly enough, as these new techniques have developed and as the scientific understanding of fast processes has improved, further progress in short pulse generation has been achieved. In 1974 the first pulses shorter than a picosecond (10^{-12} sec) were generated (Shank and Ippen, 1974) and just recently the 10^{-13} sec mark has been passed (Fork et al., 1981). As this is being written scientific investigations of a variety of high speed phenomena are being pursued into femtosecond (10^{-15} sec) domain. At the same time, compact opto-electronic devices are being developed for the purpose of making picosecond signal processing and transmission a practical reality.

The progress in short pulse generation is illustrated graphically in Figure 1.1. Passive modelocking of the Nd: glass laser (DeMaria et al., 1966) first led to the generation of pulses less than 10 psec in duration. This achievement provided much of the stimulus for picosecond technological development in the next decade. During this time, because of the high peak powers it provided, the Nd:glass also remained the work-horse of the field. A well designed system can provide transform-limited pulses in the 3-6 psec range, although repetition rates remain very low.

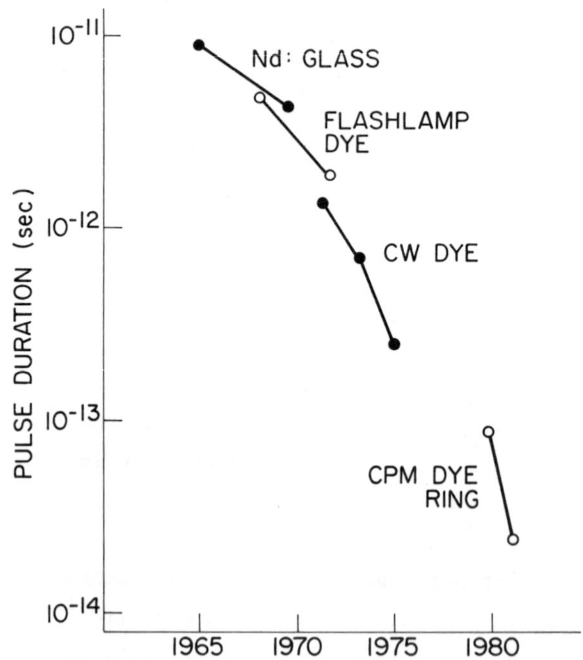

FIGURE 1.1 The decrease in optical pulsewidth as a function of time since 1965.

The passively-modelocked flashlamp-pumped dye laser (Schmidt and Schäfer, 1968; Bradley, 1977) provided both a decrease in pulsewidth and a broad extension of wavelength coverage. It was still a flashlamp system, subject to fluctuation and low repetition, but it generated pulses in a way that was fundamentally different from that of the Nd:glass laser. Streak camera investigations of ultrashort pulse formation in flashlamp-pumped dye lasers (Arthurs et al., 1973a; Arthurs et al., 1973b; Dempster et al., 1972) revealed that the mechanism of modelocking in dye lasers was different from that occuring in giant-pulse lasers. Rate equation analysis (New, 1974) confirmed that, in these lasers, the combined action of gain and absorber saturation can produce rapid pulse compression even in the absence of a fast response time in the absorber. It was this mechanism, then, that made possible the next generation of picosecond and subpicosecond sources: passively modelocked, continuously operated, dye lasers. With the first operation of such a cw system (Ippen et al., 1972) pulses as short as 1.5 psec were generated. Subsequent improvements and modifications of that system led to pulses as short as 0.3 psec (Ippen and Shank, 1975). More recently, with a bidirectional, colliding-pulse-modelocked (CPM) ring laser (Fork et al., 1981), pulse durations have moved below the 0.1 psec (100 fsec) mark. Further manipulation of these pulses by self-phase modulation in a short optical fiber and compression with a grating pair has produced pulses as short as 30 fsec (Shank et al., 1982).

Perhaps just as important as the ultrashort pulse durations are the reproducibility from pulse to pulse and the high repetition rates that are achieved with cw lasers. Powerful signal-averaging techniques can therefore be applied to picosecond and femtosecond studies and measurements can be made with low intensity pulses that do not distort the process under investigation. The decrease in "shortest" pulse duration documented in Figure 1.1 also does not indicate the variety of other new picosecond laser systems that have, during this same period, provided an ever increasing wavelength coverage and measurement versatility. We discuss some of these in Chapter 2.

As optical pulse durations from lasers have decreased dramatically during the past decade so have the capabilities of optical transmission systems expanded enormously. Optical fibers with losses as low as 0.6 dB/km near 1.3 μm and 0.2 dB/km near 1.6 μm (Mija et al., 1979) are now available. Fiber dispersion, of primary importance for short pulse propagation, can also be minimized (or optimized for special cases) in this long wavelength regime. The predicted "zero-dispersion" behavior (Li, 1978) has been confirmed by pulse-delay measurements using wavelength-tunable radiation generated in a fiber by stimulated Raman effect (Cohen and Lin, 1977; Lin et al., 1978) and by the transmission of picosecond pulses over kilometer-lengths of single mode fiber without observable pulse broadening (Bloom et al., 1979). Recent work with double-clad single mode fibers (Mija et al., 1981; Jang et al., 1982) has produced fibers with two dispersion minima and a resulting low dispersion over a broad spectral range to accomodate use at both 1.3 μm and 1.6 μm. The chromatic dispersion of such a fiber is plotted in Figure 1.2 and compared with that of a conventional

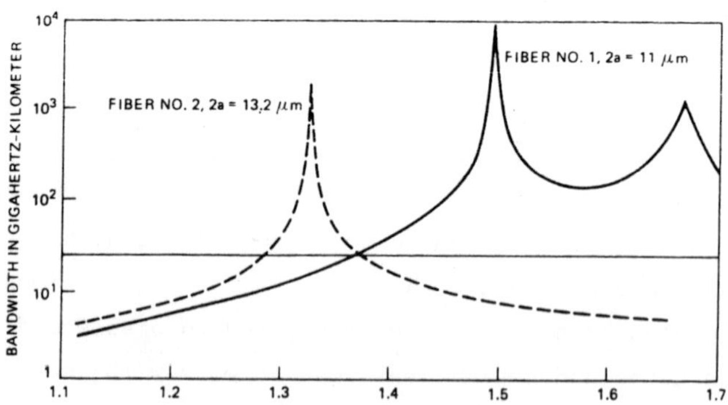

FIGURE 1.2 Transmission bandwidth of a double-clad single-mode fiber (Jang et al., 1982).

single mode fiber. From the point of view of sub-picosecond technology, it is interesting to note that within the range of 1.5 μm to 1.66 μm the dispersion of this fiber is less than 0.7 ps/km/nm. With additional **tailoring** the limit could be pushed well into the femtosecond regime.

1.2 Short Pulse Generation

The method by which short laser pulses are generated is called "modelocking" (Smith et al., 1975). That is: laser modes at different eigenfrequencies are locked together in phase to produce a coherent oscillation over the relatively broad band necessary for a short optical pulse. Although we shall see later that the most convenient treatments of short pulse formation and propagation rely on time domain analysis, a brief review of the frequency domain implications will also be helpful.

For each transverse (spatial) mode of a laser resonator there corresponds a set of longitudinal modes separated in frequency by $c/2L$, where c is the speed of light and L is the optical pathlength between the two mirrors. If the medium between the mirrors is dispersive, as it is in semiconductor diode laser for example, the spacing between resonator modes is more precisely:

$$\frac{c}{2n_g L} = \frac{c}{2L\left(n + \nu \frac{dn}{d\nu}\right)} \tag{1.1}$$

where c/n_g, the group velocity in the medium, is expressed in terms of the derivative of the index of refraction with respect to frequency ν. Note that for linear dispersion (i.e. $dn/d\nu$ = const.) the modes remain equally spaced.

Consider now the n-th mode to have amplitude E_n, angular frequency ω_n, and phase ϕ_n, then the total laser output field E_T can be written

$$E_T = \sum_n E_n \exp\{i[\omega_n(t - z/c) + \phi_n]\} + c.c. \tag{1.2}$$

where c.c. represents the complex conjugate and the observed output is travelling in the positive z direction. With equal mode spacing $\omega_n = \omega_o + n\Delta$, where $\Delta = 2\pi(c/2L)$, we can write

$$E_T = \exp\{i\omega_o(t - z/c)\} \sum_n E_n \exp\{i[n\Delta(t - z/c) + \phi_n]\} + c.c. \tag{1.3}$$

This corresponds to a carrier wave of frequency ω_o whose envelope is periodic with period $T = 2L/c$ and travels with the velocity c. For ϕ_n a constant (i.e. all phases locked together) the envelope consists of a single pulse per period with a pulsewidth given approximately by the reciprocal of the frequency range over which the E_n's have appreciable value. That is, the pulsewidth

$$t_p \simeq \{N(c/2L)\}^{-1} \tag{1.4}$$

where N is the number of oscillating modes. This frequency domain picture is illustrated in Figure 1.3. The larger the laser bandwidth, the shorter the potential pulse durations. Table I gives a rough indication of the available bandwidths in some of the most commonly mode-locked lasers. Actual pulseshapes depend upon the method of modelocking and a variety of material properties. Incomplete locking of modes results in an envelope

TABLE 1.1 Bandwidths of several common lasers

	$\Delta\nu$		$\Delta\tau$
HeNe	1 GHz		400 psec
ArgonIon	3 GHz		130 psec
Nd:YAG	10 GHz		30 psec
Nd:Glass	3,000 GHz		.25 psec
Semiconductor	10,000 GHz	(100 A°)	.05 psec
Dye	30,000 GHz	(300 A°)	.02 psec

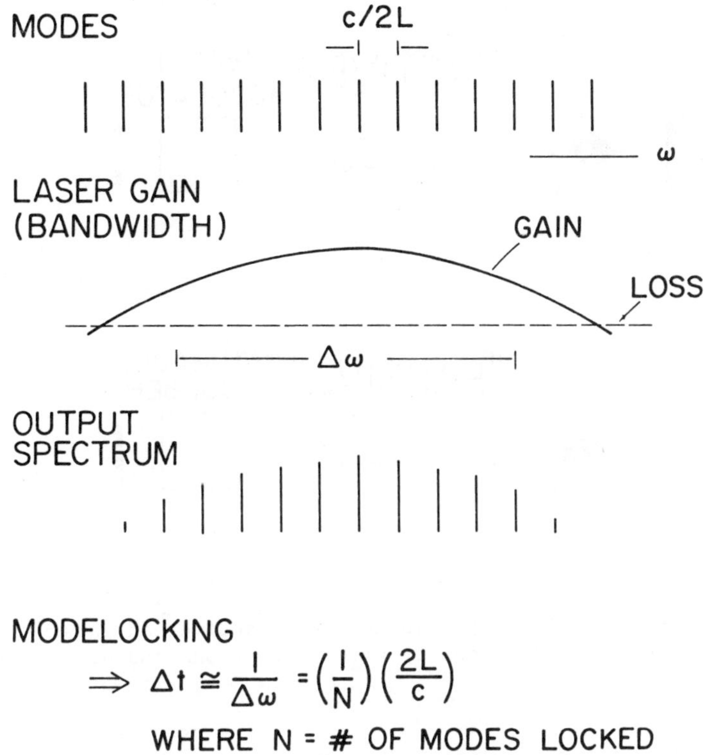

FIGURE 1.3 Oscillating spectrum and potential pulse duration of a multi-longitudinal mode laser.

larger than that predicted by the oscillating mode spectrum. The minimum pulsewidth obtainable from a given mode spectrum is referred to as the Fourier transform-limited pulsewidth. When calculating this limit one should remember that the measurable quantity, spectral density, is not the Fourier transform of the intensity but rather of the electric field autocorrelation. If a pulse is to be considered 'transform limited' the time-bandwidth product must correspond to the minimum value for its particular pulse shape (see Chapter 3).

Techniques for producing modelocking fall into two categories as illustrated in Figure 1.4, active and passive. In active modelocking, loss or phase modulation is introduced into the laser

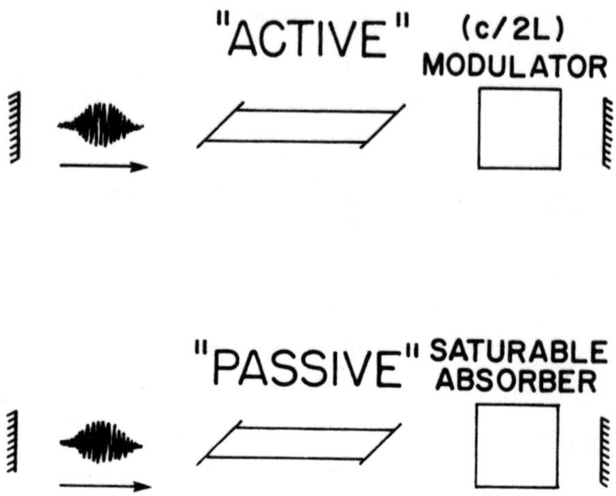

FIGURE 1.4. Schematic representations of active and passive pulse generation with an intracavity element near one end of the laser resonator.

resonator. If the modulation frequency is approximately equal to the mode frequency separation, the sidebands generated cause a phase locking to occur as illustrated in Figures 1.5 and 1.6. Such analysis can be difficult for many modes. An equally valid and more tractable approach is to consider the shaping that occurs as the pulse of light passes through the modulator each period near the point of minimum loss.

Passive modelocking involves the use of an intracavity saturable absorber that passes intense pulses of light with less loss than experienced by a constant low level of light. Under the proper conditions the laser output spontaneously breaks up into a train of "modelocked" pulses. The difference between active and passive modelocking is simply that, in the latter, the optical pulse provides its own modulation. Passive modelocking, when possible, generally results in the shortest pulses

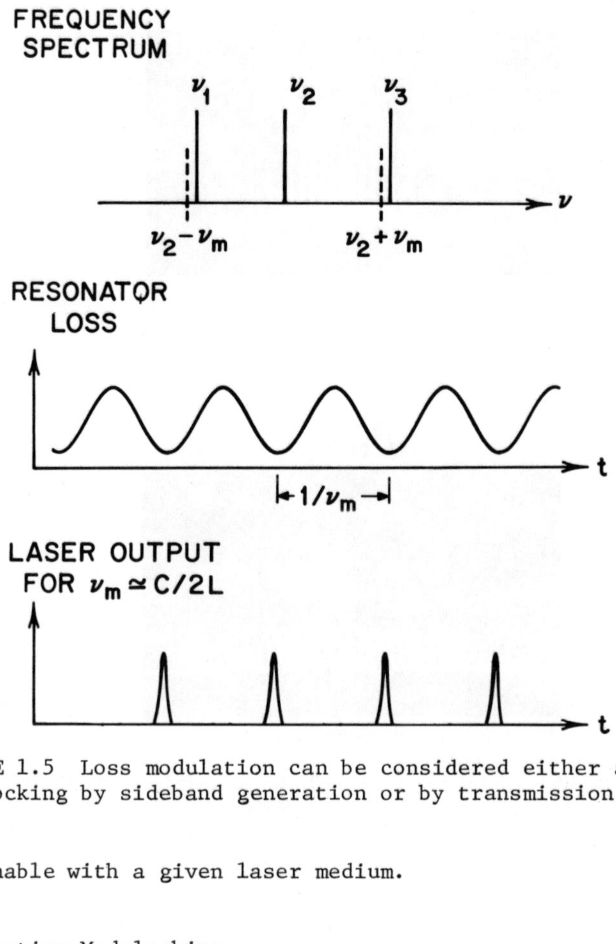

FIGURE 1.5 Loss modulation can be considered either as means of modelocking by sideband generation or by transmission gating.

obtainable with a given laser medium.

1.3 Active Modelocking

Much of the early theoretical work on modelocking focused on homogeneously broadened lasers with active, internal modulation. It was the first case to yield analytical solutions for steady state pulse shapes and pulse duration (Kuizenga and Siegman, 1970a). For homogeneous broadening and long upper-state life-time (appropriate to solid-state lasers like Nd:YAG), it can be assumed that the gain seen by a pulse travelling through the amplifier remains approximately constant in time and is given by (Yariv, 1976)

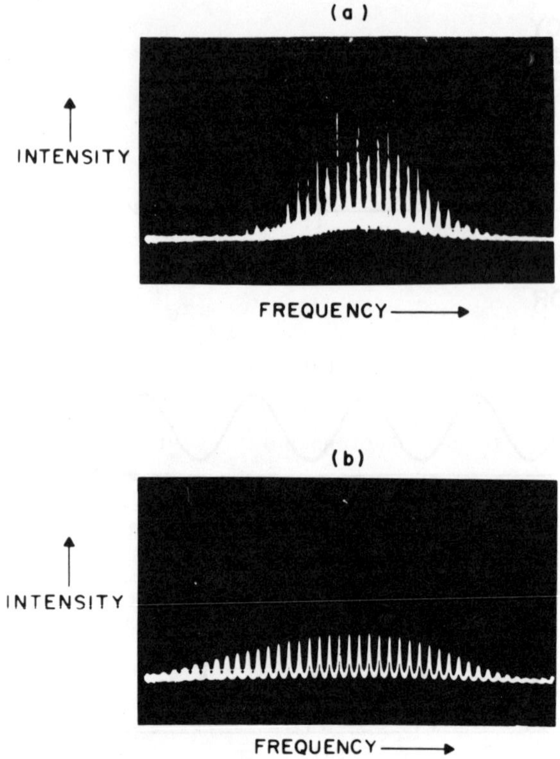

FIGURE 1.6 Frequency spectrum of (a) an unmodelocked laser and (b) the same laser after modelocking (Smith et al., 1975).

$$G = G_o(1 + I/I_s)^{-1} \qquad (1.5)$$

where I is the average internal intensity and I_s is the saturation intensity. The loss consists of two terms, a constant L and a time varying modulation $L_m(1 - \cos \omega_m t)$, which can act to shorten the pulse with each roundtrip. Pulse spreading is produced by the frequency dependence (i.e. bandwidth limitation) of the gain. If the gain profile is approximated to be parabolic in frequency about line center: $1 - [(\omega - \omega_o)/\omega_G]^2$ then its effect in the time domain can be represented by its Fourier transform equivalent: a term proportional to the second derivative of the pulse profile (Haus, 1975a; Haus, 1982). A small mismatch between modulation frequency and cavity roundtrip time, can be accounted for by addition of a term proportional to the first derivative of the pulse. One then has a differential equation for the steady state pulse amplitude (envelope) (Haus, 1982):

$$\{G - L - L_m(1 - \cos \omega_m t)\} E(t) = - \left\{\delta T_R \frac{d}{dt} + \frac{G}{\omega_G^2} \frac{d^2}{dt^2}\right\} E(t) \quad (1.6)$$

where δT_R is the detuning parameter and ω_G is the bandwidth of the gain. This is a Mathieu equation (Abramowitz and Stegun, 1972) which yields solutions periodic at ω_m. Further simplification is achieved by letting $\delta T_R = 0$ and by approximating $\cos \omega_m t$ by $1 - (\omega_m t)^2/2$. Equation (1.6) then becomes the Schroedinger equation for a particle in a parabolic potential. Its solutions are known to be of the form

$$E(t) = H_n(\omega_p t) \exp\left\{\frac{-\omega_p^2 t^2}{2}\right\} \quad (1.7)$$

where H_n is the Hermite polynomial of order n and

$$\omega_p = (\omega_m \omega_g)^{1/2} (L_m/2G)^{1/4}. \quad (1.8)$$

The full width at half maximum intensity of the lowest order pulse is

$$t_p = (2 \ln 2)^{1/2}/\omega_p. \quad (1.9)$$

It can be shown (Haus, 1975a) that gain saturation favors the lowest order mode and causes the higher order modes to be unstable in the steady state. So, pulses obtained by active modelocking of homogeneously broadened lasers with low emission cross-sections are Gaussian in shape and have a pulse duration that is inversely proportional to the square root of the gain bandwidth. These features of the theory have been verified experimentally with Nd:YAG lasers (Kuizenga and Siegman, 1970b) and with high-pressure CO_2 systems (Smith et al., 1972). One word of caution is in order regarding the above analysis. In high gain laser systems such as dyes, F-centers and semiconductors gain saturation during the pulse is appreciable and the gain cannot be assumed constant. This dynamic saturation and recovery between pulses can then lead to a shorter pulse than that predicted by (1.6).

1.4 Passive Modelocking with a Fast Saturable Absorber

The shortest pulses have been produced by passive modelocking, first demonstrated with the ruby laser (Mocker and Collins, 1965) and the Nd:glass laser (DeMaria et al., 1966) and since utilized in virtually every type of laser (Smith et al., 1975). Results obtained with passive modelocking vary considerably from system to system, depending upon the medium response times as well as gain and absorber cross-sections. It is convenient for purposes of analysis to separate saturable absorbers into two classes: fast (or hard) and slow (or soft). A fast saturable absorber is one that recovers on a time scale shorter than the pulse duration so that it represents a loss given by

$$L_A = L_o \left(1 + \frac{I(t)}{I_{SA}}\right)^{-1} \qquad (1.10)$$

where I_{SA} is the saturation intensity of the absorber and L_o is the small signal loss. In a dye, $L_o = \sigma_A N\ell$ where σ_A is the absorber cross-section, N is the concentration and ℓ is the path length. The saturation intensity is $\hbar\omega_o/\sigma_A\tau$ where τ is the dye absorption recovery time.

The steady state theory of fast absorber modelocking can be discussed in terms of an electronic regenerative pulse generator (Cutler, 1955). With a nonlinear element, called an expandor, in the feedback loop, Cutler showed that such a generator produced pulses with separation equal to the group delay of one roundtrip and with a duration inversely proportional to the bandwidth. Cutler's analysis even provided expressions for pulse duration in the presence of group velocity dispersion in the loop.

If we expand the saturable loss to first order in $I(t) = |E(t)|^2$ such that

$$L_A \simeq L_o \left\{1 - \frac{|E(t)|^2}{I_{SA}}\right\} \tag{1.11}$$

we can again write a differential equation for the steady pulse envelope (Haus, 1975b):

$$\{G - L\}\, E(t) = - \left\{\frac{G}{\omega_G^2}\frac{d^2}{dt^2} + \frac{L_o|E(t)|^2}{I_{SA}}\right\} E(t) \tag{1.12}$$

for which the gain, loss, and bandwidth limiting terms have the same meaning as in the previous section on active modelocking. Here again we note that the gain plays no dynamic role in pulse shaping. All of the self-narrowing comes from the instantaneously responding saturable absorption term. This self-narrowing achieves a balance with the spreading due to limited gain bandwidth. The solution is

$$E(t) = E_o \operatorname{sech}(t/t_p) \tag{1.13}$$

where

$$t_p = \frac{1}{\omega_G}\left\{\frac{2GI_{SA}}{L_o|E_o|^2}\right\} \tag{1.14}$$

and

$$L - G = \frac{G}{\omega_G^2 t_p^2}\ . \tag{1.15}$$

Several features are worth mentioning. The pulsewidth is now inversely proportional to the bandwidth, and the wings of the pulses are exponential rather than the Gaussian characteristic of active modelocking.

Comparison of experiment with theory in this case is somewhat more difficult. Lasers for which fast saturable absorber modelocking is important (i.e. Nd:glass and Nd:YAG) tend to be operated in a transient rather than steady state regime. By the time steady state is reached, pulses in the laser can be so intense that nonlinear effects give rise to spectral broadening and pulse distortion (von der Linde, 1972; Duguay et al., 1970; Korobkin et al., 1970). For most scientific applications, therefore, single pulses are selected from the pulse train during transient build-up (von der Linde, 1972; Zinth et al., 1977). The selection and evolution of picosecond pulses from noise fluctuations has been analyzed in considerable detail theoretically and numerically (Fleck, 1970; Kryukov and Letokhov, 1972; New, 1979). Various experimental studies with different systems have found pulses approximately $\mathrm{sech}\,(t/t_p)$ in shape (Auston, 1971), pulses with Gaussian leading edges but exponential tails (von der Linde, 1972) and well-defined Gaussian pulses (von der Linde and Laubereau, 1971). One final, practical point is that the saturable absorbers used in these lasers are only "fast" in a relative sense. In fact, they have recovery times in the range 8-9 psec (von der Linde and Rodgers, 1973) which may work well to select pulses but which is not shorter than the pulses in most cases. This will certainly introduce some asymmetry into the shaping process.

1.5 Passive Modelocking with a Slow Saturable Absorber

Studies of passively modelocked flashlamp-pumped dye lasers (Arthurs et al., 1973a, 1973b; Dempster et al., 1972) revealed that the saturable absorber recovery times in these lasers were in the range 100 psec to 1 nsec, much longer than the duration of the pulses being generated. An insightful rate equation analysis by New (1974) then confirmed that the combined action

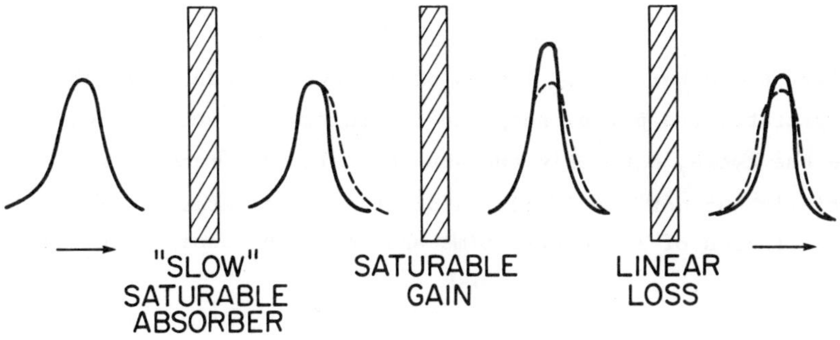

FIGURE 1.7 Transmission of a pulse through a slow saturable absorber, saturable amplifier, and linear loss can result in pulse shortening.

of gain saturation and absorber saturation could generate ultrashort pulses even in the absence of a fast recovery time. The idea is illustrated in Figure 1.7. As the pulse passes through the slow absorber, its front edge is absorbed more strongly than the peak and the trailing edge. The absorber simply becomes increasingly more transmissive with time since its recovery time is longer than the pulse duration. A simple dye rate equation in this limit yields:

$$L(t) = L_o \exp\left\{-\frac{\sigma_A}{\hbar\omega_o}\int_{-\infty}^{t} |E(t)|^2\, dt\right\} \tag{1.16}$$

where $|E(t)|^2$ is again the optical intensity in watts/cm^2, $\hbar\omega_o$ is the photon energy, σ_A is the absorber cross-section, and $L_o = \sigma_A N\ell$ is the small signal absorption.

When the pulse passes through the gain medium it receives less amplification in the trailing edge because of gain saturation. The gain, which also recovers slowly on the time scale of the pulse, is given by

$$G(t) = G_o \exp\left\{-\frac{\sigma_G}{\hbar\omega_o}\int_{-\infty}^{t} |E(t)|^2\, dt\right\} . \qquad (1.17)$$

Finally, the other (D.C.) losses of the system reduce, in the steady state, the pulse energy to its initial value. To complete the cycle, both gain and absorber elements recover substantially but not completely before the pulse returns. Within certain ranges of round trip time and under the basic condition

$$\sigma_A > \sigma_G \qquad (1.18)$$

it can be shown (New, 1974; Haus, 1975c) that a pulse of constant (steady state) energy can be progressively shortened with each round trip. Although an excess of gain is needed to start pulse build-up, in the steady state the absorption is at a higher value than the gain until arrival of the pulse. The greater saturability of the absorber results in the peak of the pulse receiving net gain; then gain saturation leads to net loss for the trailing edge. Figure 1.8 illustrates this behavior as a function of time.

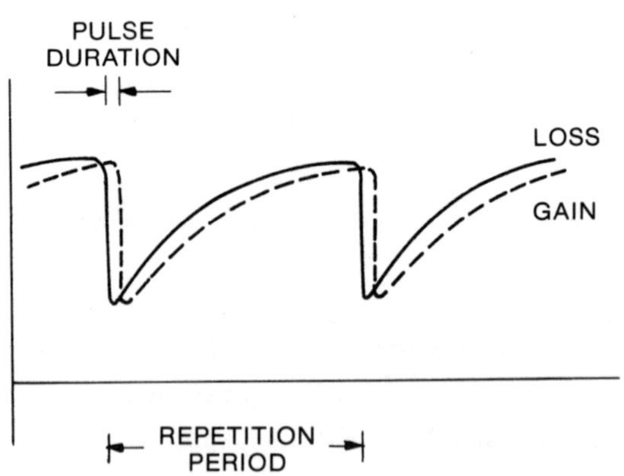

FIGURE 1.8 Steady state dynamics of a laser modelocked by a slow saturable absorber. Note that the gain recovers less rapidly than the loss so that the leading edge of the pulse experiences loss.

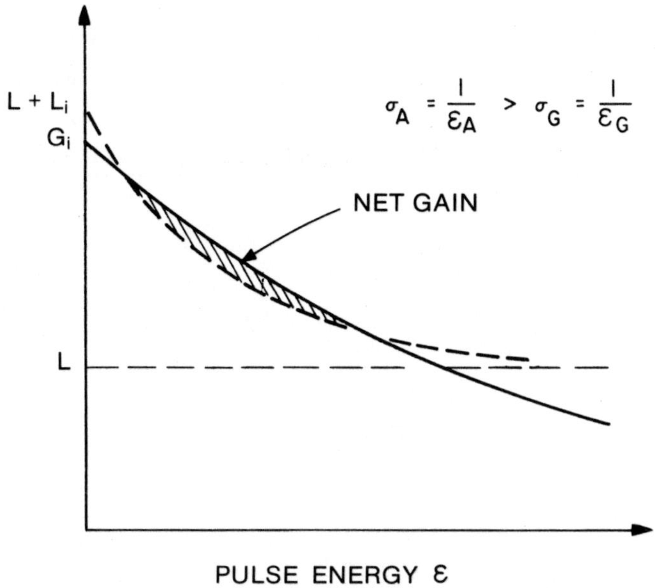

FIGURE 1.9 Gain and loss as a function of pulse energy.

Figure 1.9 illustrates the saturation of both gain and loss according to (1.16) and (1.17) and the inequality of (1.18). The integrated pulse energy ε advances in time from zero, where there is net loss, through the region of net gain to the end of the pulse where the gain is again less than the loss. In (1.16) and (1.17) we have used the same pulse intensity $|E(t)|^2$. In practice it is common to focus more tightly in the saturable absorber. In our notation this can be accounted for by an "effective" σ_A which is increased relative to σ_G by the inverse ratio of the beam areas.

With certain approximations we can again write a differential equation for this case (Haus, 1975c; Haus, 1982):

$$\{G(t) - L(t) - \ell\} E(t) = -\left\{\frac{L}{\omega_G^2}\frac{d^2}{dt^2} + \frac{\delta}{\omega_G}\frac{d}{dt}\right\} E(t) \qquad (1.19)$$

where the gain $G(t)$ and the loss $L(t)$ have the dependence on pulse energy discussed above. A detuning term (proportional to d/dt) must be added here to facilitate a solution. Remember that for active modelocking and for modelocking with a fast absorber it was possible to find a solution with zero detuning, but that is not the case here. To obtain an analytical solution we expand (1.16) and (1.17) to second order in pulse energy

$$G(t) \simeq G_i \left\{1 - \sigma_G\, \varepsilon(t) + \frac{\sigma_G{}^2 \varepsilon^2}{2}(t)\right\} \tag{1.20}$$

and

$$L(t) \simeq L_i \left\{1 - \sigma_A \varepsilon(t) + \frac{\sigma_A{}^2 \varepsilon^2}{2}(t)\right\} \tag{1.21}$$

where

$$\varepsilon(t) = \frac{1}{\hbar\omega_o} \int_{-\infty}^{t} |E(t)|^2\, dt \tag{1.22}$$

and G_i and L_i are the "initial" values of gain and loss just prior to pulse arrival. Both G_i and L_i depend upon recovery times and cavity round-trip time.

The solution of (1.19) is

$$E(t) = E_o\, \mathrm{sech}(t/t_p) \tag{1.23}$$

where

$$t_p = \frac{4}{\omega_G} \left(\frac{1}{\sigma_A \varepsilon_\infty}\right) (L/L_i)^{1/2}. \tag{1.24}$$

Here ε_∞ is the total pulse energy (i.e. $t \to \infty$ in (1.22)) and L/L_i is the ratio of D.C. loss to saturable absorber loss. The pulsewidth is again proportional to the inverse of the resonator bandwidth which we have called the gain bandwidth. If the solution (1.23) is plugged back into (1.19), one gets, in addition to (1.24), two other characteristic equations for the pulse amplitude E_o and the detuning parameter δ.

This analysis by Haus (1975c) provided the first closed form solutions of pulse shape and the first predictions of pulse duration in the presence of bandwidth limitation. Although several approximations are made, namely parabolicity of the frequency restriction and second-order expansion of loss and gain, the results obtained from this theory have been found to be surprisingly accurate when compared with more involved numerical calculations (Hermann et al., 1981; Hermann and Weidner, 1982; Stix and Ippen, 1982).

With successful operation recently of the passively mode-locked dye laser in a bidirectional (CPM) ring configuration (Fork et al., 1981), some interest has been focused on the effect of coherent pulse coupling and interation in the saturable absorber. The modelocking dynamics of this system are similar in many respects to those of conventional slow absorber mode-locking. Important differences are: (a) pulses traveling in opposite directions pass through the absorber at the same time (if the absorber is thinner than the pulses) but through the gain medium at different, well-separated times; and (b) the colliding pulses in the absorber interfere to form an absorption grating which in turn coherently scatters light from each pulse into the other. The first condition results in an effective increase in the ratio of absorber cross section σ_A to gain cross section σ_B by a factor of 2. This provides increased stability and can help generate shorter pulses. The coherent coupling interaction (b) can be included in the analysis numerically (Hermann et al., 1981; Stix and Ippen, 1982). The bleaching and generation of an absorption grating as a function of time in the thin absorber are illustrated in Figure 1.10. Pulse shaping due to these dynamics are given in Figure 1.11 for a single pass through the absorber. The dash-dot line is the shape of the input pulse; the dashed shape would be the transmitted shape of one of the pulses if it passed through the absorber by itself; the dotted line is also for a single pulse,

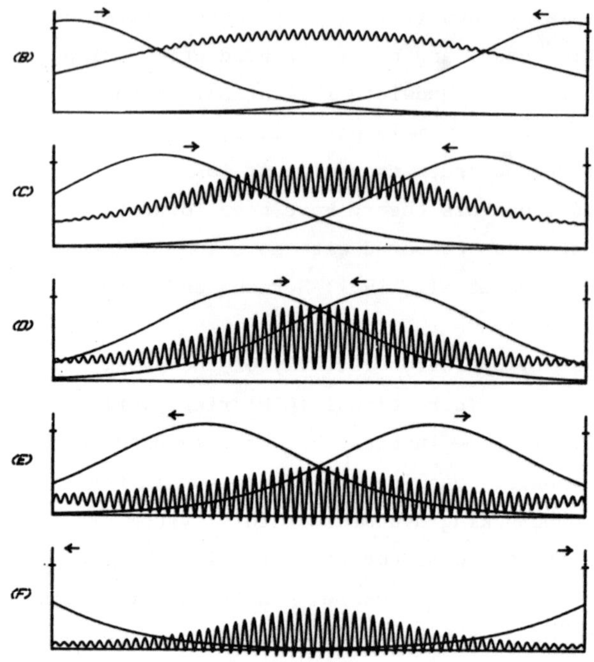

FIGURE 1.10 Pulses incident from opposite ends interface and create an absorption grating. Time sequence from top to bottom.

but focused to one-third the area; and the solid line is the actual result obtained with colliding pulses and inclusion of the grating effect. Surprisingly, the complication introduced by coherent coupling can very closely be mimicked by tighter focusing (or by an increase in the absorber cross-section). The enhancement due to the coupling is a factor of 1.5, so that the overall enhancement in CPM is the factor of three. Although Figure 1.11 shows only one pass through the saturable absorber, similar results have been obtained for steady state pulses in a system with saturable loss; saturable gain, and dispersion. As the pulses begin to get shorter than the thickness of the absorber, the cross-section enhancement factor, and hence the pulse

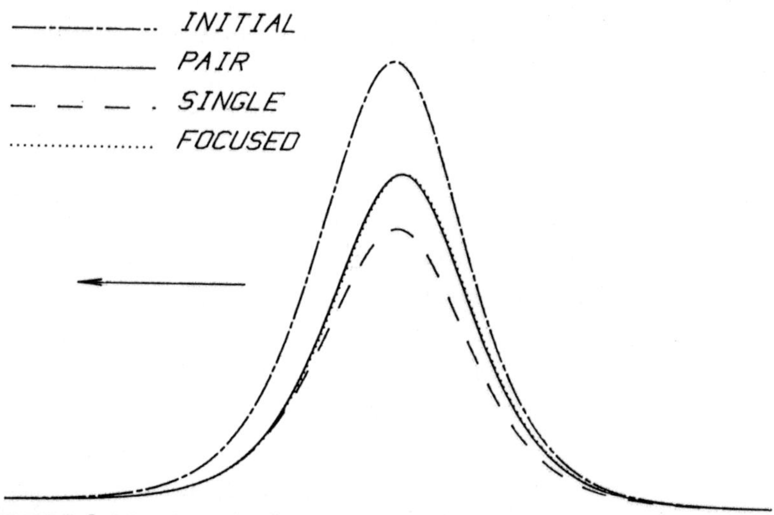

FIGURE 1.11 Numerical computed pulse shapes before and after passage through a slow saturable absorber (see text).

shortening velocity, begins to decrease, but pulses do continue to get shorter until a balance is struck with the dispersion of the resonator. At the current, femtosecond, state-of-the-art, small amounts of dispersion in dye streams and in mirror coatings become increasingly problematic. Schemes for generating shorter pulses will most likely involve clever, dispersion-compensating design of these passive elements. An alternative approach is the nonlinear optical compression described below.

1.6 Ultrashort Pulse Propagation

In the previous sections we discussed ultrashort pulse generation by nonlinearly absorbing and amplifying elements. Now we turn to the propagation of such ultrashort pulses through essentially transparent media of also essentially unlimited bandwidth (in the spectral amplitude sense we have used above). In such media, and optical fibers are good examples, the primary forces of pulse distortion are group velocity dispersion and nonlinear phase changes. We now write the wave equation for the electric field amplitude in the form (Whitham, 1974; Haus, 1982)

$$\frac{\partial E}{\partial z} + \frac{1}{v_g}\frac{\partial E}{\partial t} = \frac{i}{2}\frac{\partial^2 \beta}{\partial \omega^2}\frac{\partial^2 E}{\partial t^2} - \frac{i\pi}{\lambda n} n_2 |E|^2 E. \tag{1.25}$$

The left hand side of this equation describes free propagation of the pulse with the group velocity $v_g = (\partial\beta/\partial\omega)^{-1}$. The usual, slowly-varying-envelope approximation has been assumed. On the right hand side of the equation, the first term is a diffusion term arising from variation of the group velocity with frequency; the second term can be thought of either as a nonlinear velocity term or as a nonlinear (self-phase modulation) driving polarization. With a rather simple transformation, (1.25) reduces to the nonlinear Schroedinger equation (Mollenauer et al., 1980; Nakatsuka et al., 1981; Haus, 1982).

Let us first consider the distorting terms separately. After propagation over a long distance a pulse will spread out due to group velocity dispersion with (for normal dispersion $\partial^2\beta/\partial\omega^2 < 0$) the high frequencies trailing the lower frequencies. It can be shown (Unger, 1977; Marcuse, 1980) that a coherent Gaussian pulse of duration (full $1/e$ intensity width) τ will broaden to a width

$$\tau' = \tau\sqrt{1 + (\tau_c/\tau)^4} \tag{1.26}$$

where the 'critical" pulsewidth is

$$\tau_c = (2)^{5/4}\sqrt{L\,\frac{\partial^2\beta}{\partial\omega^2}}$$

for a propagation length L. Clearly pulses shorter than τ_c spread rapidly.

If a pulse is dispersed in the above manner, no spectral amplitudes have been changed and the pulse can be reconstructed by transmission through an equivalent length of opposite dispersion, as has been suggested for optical fiber systems (Marcuse and Lin,

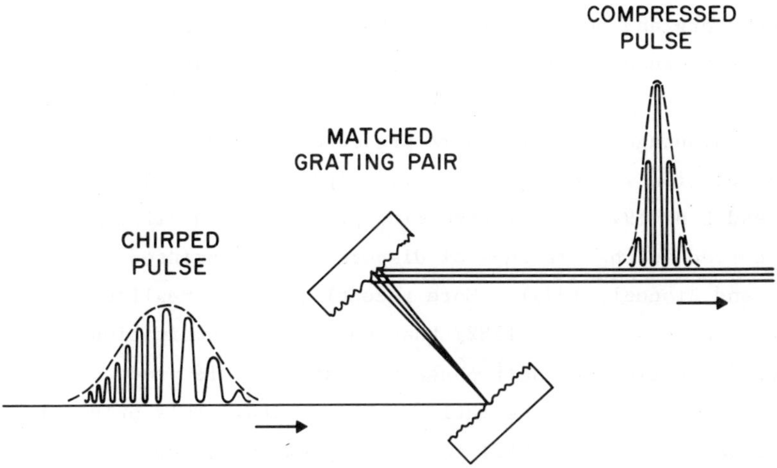

FIGURE 1.12 Compression of a chirped pulse by a matched pair of gratings (Treacy, 1969).

1981). A laboratory technique for this recompression of a chirped pulse is shown in Figure 1.12 (Treacy, 1969). The grating pair technique is versatile because the degree of anomalous dispersion is easily variable by changing the distance L between the gratings

$$L \frac{\partial^2 \beta}{\partial \omega^2} = \frac{L\lambda^3}{2\pi c^2 d^2 \{1 - (\lambda d - \sin \gamma)^2\}} \tag{1.28}$$

where d is the number of lines/cm on the grating and γ is the angle between the incident light and the grating normal. Recompression has also recently been demonstrated in sodium vapor (Nakatsuka et al., 1981) and with excitonic dispersion in GaAs (Fehrenbach and Salour, 1982).

Through the self-phase modulation term on the right hand side of (1.25) a pulse can cause a rapid modulation of its own phase and hence a frequency upshift on the rise and a downshift in the trailing half. Since the pulse shape is not changed by this action, anamolous dispersion, such as that provided by a grating pair, may then be used to compress the pulse (Laubereau,

1969; Lehmberg and McMahon, 1976). Pulse compression is limited in general because the frequency chirp is not linear over the entire pulse.

The advantage optical fibers presented for enhancing self-phase modulation was recognized early (Ippen et al., 1974; Stolen and Lin, 1978) and interesting pulse shaping effects that occurred in the presence of dispersion were pointed out (Fisher and Bischel, 1975). More recently, it was realized (Grischkowsky and Balant, 1982) that the proper combination of fiber dispersion and self-phase modulation could produce a more uniform chirp over the entire pulse length. This principle then led (Shank et al., 1982) to the compression of pulses to 30 fsec with a 15 cm fiber and a grating pair.

A number of years ago it was also pointed out (Hasegawa and Tappert, 1973) that the self-phase modulation term could be used to compensate for the pulse broadening effect of dispersion. In the simplest case, with the proper sign of dispersion and with a $\mathrm{sech}(t/t_p)$ pulse envelope, the two terms on the right hand side of (1.25) can cancel. The pulse then propagates distortion free as a 'soliton'. Higher order soliton solutions that are periodic in their propagation characteristics are also possible (Zakharov and Shabat, 1973; Satsuma and Yajima, 1974). With the advent of picosecond laser sources (Mollenauer and Bloom, 1979) in the wavelength regime of negative group velocity dispersion ($\sim 1.55\ \mu m$), experimental demonstration of soliton formation and propagation was made possible (Mollenauer et al., 1980). In addition to opening up the possibility of interesting, scientific studies of soliton behavior, this work may point the way to eventual, practical control of ultrashort pulse propagation in optical fibers.

REFERENCES

Abramowitz, H., and I.A. Stegun, eds., 1972, *Handbook of Mathematical Functions* (Dover, New York).

Alfano, R.R., and S.L. Shapiro, 1971, Chem. Phys. Lett. 3, 407.

Arthurs, E.G., D.J. Bradley, and A.G. Roddie, 1973a, Chem. Phys. Lett. 22, 230.

Arthurs, E.G., D.J. Bradley, and A.G. Roddie, 1973b, Appl. Phys. Lett. 23, 88.

Auston, D.H., 1971, Appl. Phys. Lett. 18, 249.

Bloom, D.M., L.F. Mollenauer, C. Lin, D.W. Taylor, and A.M. DelGaudio, 1979, Opt. Lett. 4, 297.

Bradley, D.J., 1977, *Ultrasound Light Pulses*, Topics in Appl. Phys. 18, edited by S.L. Shapiro (Springer Verlag).

Cohen, L.G., and C. Lin, 1977, Appl. Opt. 16, 3136.

Cutler, C.C., 1955, Proc. I.R.E. 43, 140.

DeMaria, A.J., D.A. Stetser, and H. Heyman, 1966, Appl. Phys. Lett. 8, 22.

Dempster, D.N., T. Morrow, R. Rankin, and G.F. Thompson, 1972, J. Chem. Soc. Faraday II 68, 1479.

Edgerton, H.E., 1932, J. SMPTE 18, 356.

Fehrenbach, G.W., and M.M. Salour, 1982, *Picosecond Phenomena III* (Springer-Verlag).

Fisher, R.A., and W.K. Bischel, 1975, J. Appl. Phys. 46, 4921.

Fork, R.L., B.I. Greene, and C.V. Shank, 1981, Appl. Phys. Lett. 38, 671.

Grishkowsky, D., and A.C. Balant, 1982, Appl. Phys. Lett.

Haus, H.A., 1975a, IEEE J. Quant. Electron. QE-11, 323.

Haus, H.A., 1975b, J. Appl. Phys. 46, 3049.

Haus, H.A., 1975c, IEEE J. Quant. Electron. QE-11, 736.

Haus, H.A., 1982, *Waves and Fields in Optoelectronics* (Prentice Hall).

Hermann, J., and F. Weidner, 1982, to be published.

Hermann, J., F. Weidner, and B. Wilhelmi, 1981, Appl. Phys. B 26, 147.

Ippen, E.P., and C.V. Shank, 1975a, Appl. Phys. Lett. 27, 488.

Ippen, E.P., C.V. Shank, and A. Dienes, 1972, Appl. Phys. Lett. 21, 348.

Ippen, E.P., C.V. Shank, and T.K. Gustafson, 1974, Appl. Phys. Lett. 24, 190.

Jang, S.J., L.G. Cohen, W.L. Mammel, and M.A. Saifi, 1982, B.S.T.J. 61, 385.

Korobkin, V.A., A.A. Malyutin, and A.M. Prokhorov, 1970, JETP Lett. 12, 150.

Kuizenga, D.J., and A.E. Siegman, 1970a, IEEE J. Quant. Electron. QE-6, 694.

Kuizenga, D.J., and A.E. Siegman, 1970b, IEEE J. Quant. Electron. QE-6, 709.

Kryukov, P.F., and V.S. Letokhov, 1972, IEEE J. Quant. Electron. QE-8, 766.

Laubereau, A., 1969, Phys. Lett. 29A, 539.

Lehmberg, R.H., and J.M. McMahon, 1976, Appl. Phys. Lett. 28, 204.

Li, T., 1978, IEEE Trans. Commun. 26, 946.

Lin, C., L.G. Cohen, W.G. French, and V.A. Foertmeyer, 1978, Electron. Lett. 14, 170.

Marcuse, D., 1980, Appl. Opt. 19, 1653.

Marcuse, D., and C. Lin, 1981, IEEE J. Quant. Electron. QE-17, 869.

Mija, T., Y. Terunuma, T. Hosaka, and T. Miyashita, 1979, Electron. Lett. 15, 106.

Mija, T., K. Okamoto, Y. Ohmori, and Y. Sasaki, 1981, IEEE J. Quant. Electron. QE-17, 858.

Mocker, H.W., and R.J. Collins, 1965, Appl. Phys. Lett. 7, 270.

Mollenauer, L.F., R.H. Stolen, and J.P. Gordon, 1980, Phys. Rev. Lett. 45, 1095.

Nakatsuka, H., D. Grishkowsky, and A.C. Balant, 1981, Phys. Rev. Lett. 47, 910.

New, G.H.C., 1974, IEEE J. Quant. Electron. QE-10, 115.

New, G.H.C., 1979, Proc. IEEE 67, 380.

Porter, G., 1950, Proc. Roy. Soc. (London) A200, 284.

Satsuma, J., and N. Yajima, 1974, Supp. Prog. Theor. Phys. 55, 284.

Schmidt, W., and F.P. Schäfer, 1968, Phys. Lett. 26A, 558.

Shank, C.V., and E.P. Ippen, 1974, Appl. Phys. Lett. 24, 373.

Shank, C.V., R.L. Fork, R. Yen, R.H. Stolen, and W.J. Tomlinson, 1982, Appl. Phys. Lett. 40, 761.

Shapiro, S.L., 1977, *Ultrashort Light Pulses*, Topics in Appl. Phys. 18, edited by S.L. Shapiro (Springer Verlag), 1-15.

Smith, P.W., M.A. Duguay, and E.P. Ippen, 1975, Prog. in Quant. Electron., edited by Sanders and Stenholm (Pergamon Press), 8, 107-229.

Stix, M., and E.P. Ippen, 1982, to be published.

Stolen, R.H., and C. Lin, 1978, Phys. Rev. A17, 1448.

Treacy, E.B., 1969, IEEE J. Quant. Electron. QE-5, 454.

Unger, H.G., 1977, Arch. Elec. Ubertragung 31, 518.

von der Linde, D., 1972, IEEE J. Quant. Electron. QE-8, 328.

von der Linde, D., and A. Laubereau, 1971, Opt. Commun. 3, 27.

von der Linde, D., and K.F. Rodgers, 1973, IEEE J. Quant. Electron. QE-9, 960.

von der Linde, D., O. Bernecker, and A. Laubereau, 1970, Opt. Commun. 2, 215.

Whitham, G.B., 1974, *Linear and Nonlinear Waves* (John Wiley and Sons, New York).

Yariv, A., 1976, *Introduction to Optical Electronics* (Holt, Rinehart, Winston).

Zakharov, V.E., and A.B. Shabat, 1973, Sov. Phys. JETP 37, 823.

Zinth, W., A. Laubereau, and W. Kaiser, 1977, Opt. Commun. 22, 161.

2. PICOSECOND AND SUBPICOSECOND LASER SYSTEMS

2.1 Introduction

Although the first picosecond pulses were generated more than fifteen years ago (DeMaria et al., 1966), we are now witnessing a rapidly increasing interest in ultra-short pulse techniques and their application to dynamic studies. This is largely due to the emergence of new laser systems that offer broader wavelength coverage, sub-picosecond resolution, more sensitivity, high repetition rates, and greater reliability. Figure 2.1 compares, in two dimensions at least, several of the most important ultrashort pulse laser systems currently available. The dots indicate the primary oscillating wavelengths and easily generated harmonics of the Nd:glass and cw passively-modelocked Rhodamine 6G systems. Connecting lines indicate that all intermediate wavelengths have been generated by nonlinear optical, frequency conversion techniques. As indicated by the bars, synchronously-pumped dye lasers and color center lasers offer continuously tunable cw pulse trains, albeit not with a single pumping source or gain medium. It should

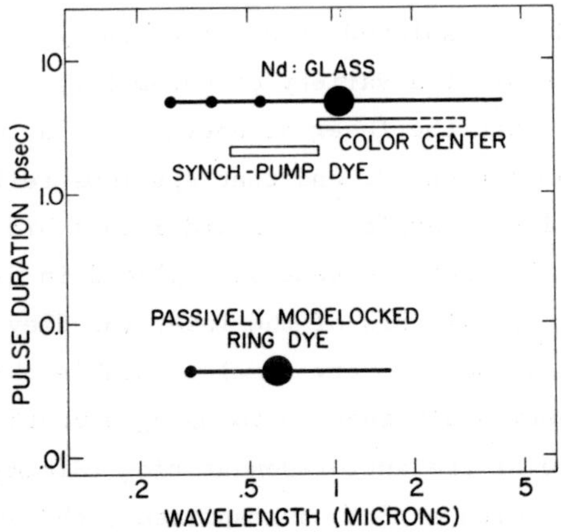

FIGURE 2.1 Wavelength coverage and typical pulse durations of available ultrashort pulse lasers.

be noted that specific experimental considerations influence the choice of system for a particular application; pulse duration may not be the only factor determining temporal resolution in each case. Nevertheless, the passively-modelocked cw dye laser has a clear advantage in this regard. It is also a high repetition rate system that can take advantage of powerful signal averaging techniques.

In this chapter we review and discuss some of the characteristics of the systems indicated in Figure 2.1.

2.2 Nd:Glass Laser

Over the years the pulse output characteristics of this laser have perhaps been the most extensively investigated (Shapiro, 1977). These investigations have to a certain extent been necessitated by the variety and variability of pulse outputs obtained. Pulse characteristics vary greatly with differences in laser design, from shot to shot in a particular system, and with position in the train of pulses produced by a single flash. Much of this behavior can be explained in terms of transient laser build-up from an initially statistical distribution of fluctuations (Kryukov and Letokhov, 1972). Selection of a single, short, isolated pulse under these conditions requires careful control of a variety of parameters. It is especially important that the laser be operated in a low order transverse mode, near threshold, and that spurious reflections and beam irregularities be avoided. In order to avoid satellite pulses the saturable absorber is generally placed in direct contact with one of the end mirrors of the laser (Bradley et al., 1970). A fast (picosecond recovery time) saturable absorber is required. An interesting alternative to using a contacted dye cell involves the use of the anti-resonant ring (Siegman, 1973) shown in Figure 2.2. With a 50% beamsplitter, the antiresonant ring has the property that all of the incident light is reflected back along the incoming path. Pulses traveling in opposite di-

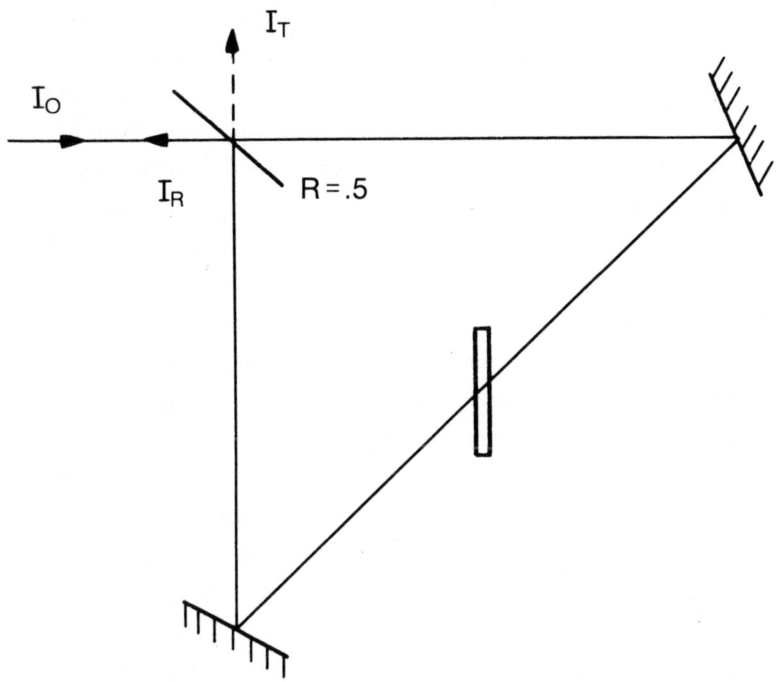

FIGURE 2.2 An antiresonant ring configuration for effective saturable absorber modelocking

rections around the ring always have the same path lengths and exactly cancel in the transmission direction. A saturable absorber positioned where the counterpropagating pulses collide is effectively at the end of the resonator. It can be inserted at Brewster's angle to reduce the usual etalon effects of a contacted dye cell. Pulsewidth reduction from 30 psec to 15 psec in a passively modelocked Nd:YAG laser has recently been obtained with this scheme (Vanherzeele and Siegman, 1982).

The build-up of Nd:glass laser pulse trains and the variation of pulse characteristics during this build-up have been studied by many workers (Shapiro, 1977). During the first part of the flashlamp pumping, spontaneous emission fluctuations are amplified linearly. When they reach sufficient intensity to begin bleaching the saturable absorber, the largest fluctuation peak grows most rapidly and takes over. Although gain saturation plays a role in emphasizing the largest peak (Glenn, 1975) it probably does not have a dramatic effect on pulse shape. (This is in direct contrast to dye lasers where strong gain depletion and gain recovery can occur within a cavity round-trip time.) As the main intensity peak (now a "mode-locked pulse") continues to grow in amplitude, nonlinearities of the glass laser medium (self-phase modulation and self-focusing) start to introduce new spectral broadening and temporal sub-structure (Duguay et al., 1970; Korobkin et al., 1970). If the gain is too high, the pulse can literally be torn apart (Eckhardt et al., 1974).

Because of this great variation between pulses within a train, accurate picosecond measurements must be made with single, properly selected, pulses (von der Linde, 1972; Zinth et al., 1977). Pulse selection is accomplished by passing the pulse train through a high speed Pockels cell or Kerr shutter that is activated by a high voltage spark gap (von der Linde et al., 1970). When the laser output reaches a predetermined level, the spark gap fires and the next pulse is transmitted. Usually the pulse is chosen to be near the beginning of the train (as viewed on an oscilloscope). At this point the laser is already in the regime where the primary fluctuation peak has been isolated but subsequent broadening has not yet come strongly into play. In a carefully designed system it is possible, in this way, to obtain pulses with an average pulsewidth of about 5 psec and an rms width fluctuation of about 0.5 psec (Zinth et al., 1977). These pulses can be in a low order transverse mode and have energies of several millijoules. Thus peak powers are around 10^9 watts, already sufficient for efficiently driving a variety of nonlinear processes. In nonlinear experiments, however, the fluctuation of pulse characteristics becomes increasingly problematic. Care must be taken to characterize each pulse used.

Some applications require even higher powers than can be obtained by a single Nd:glass oscillator. Just a few comments are in order here. Picosecond pulses saturate amplifier gain with pulse energy. In the case of Nd:glass, saturation energy density is about $5J/cm^2$. This means that Nd:glass amplifiers can only be operated in a small-signal-gain, low efficiency configuration. Even with 5 psec pulses, operation near saturation would result in peak powers of 10^{12} $watts/cm^2$. At this point destructive nonlinear effects in the glass would be overwhelming. Although pulse energies of several joules can be achieved, such very high power systems are complicated, expensive, and are needed only for a very few applications such as very short wavelength generation (Reintjes et al., 1977; Attwood et al., 1977).

An area in which Nd:glass systems have been especially useful has been in the generation and utilization of picosecond infrared pulses (Kaiser et al., 1980; Kaiser and Laubereau, 1977). With $9400\ cm^{-1}$(1.06 μm) pulses it is possible to generate tunable infrared pulses between $2500\ cm^{-1}$ and $7000\ cm^{-1}$ by three-photon parametric mixing in $LiNbO_3$ (Laubereau et al., 1974). With two $LiNbO_3$ crystals, separated by about 50 cm, considerable reduction in divergence and bandwidth is achieved (Seilmeier et al., 1978). By addition of yet another stage, and by introducing appropriate delay between the pump and signal beam, pulse durations have reduced into the sub-picosecond regime (Fendt et al., 1979). Generation of broadband tunable picosecond pulses for spectroscopy in the 10 μm - 20 μm by difference frequency mixing in CdSe has also been reported (Campillo et al., 1979).

2.3 Synchronously Pumped CW Dye Lasers

The most common picosecond laser systems at this time are commercially available, synchronously pumped dye lasers. They offer tunable cw trains of picosecond pulses at wavelengths throughout the visible and into the near infrared. An actively

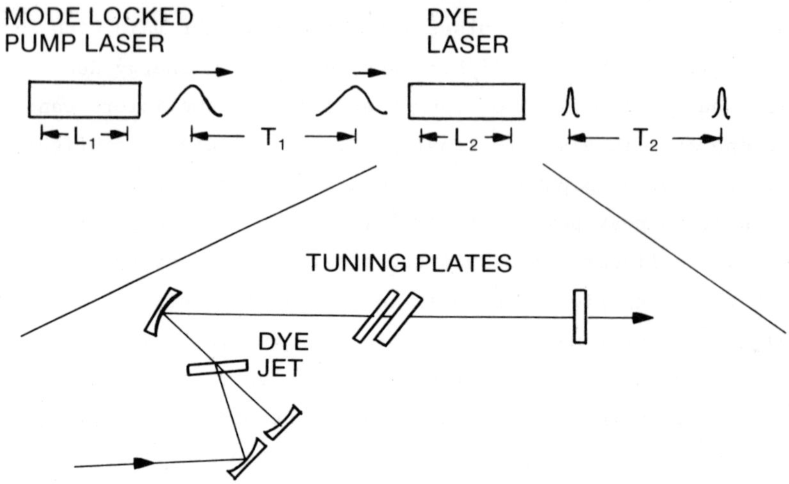

FIGURE 2.3 Schematic of a typical synchronously-pumped laser. The dye is contained in a free-flowing jet and birefringent plates provide tuning and bandwidth limitation.

modelocked pump laser (usually either an argon or krypton ion laser) is used to pump a cw dye laser of matching cavity length. A typical arrangement is shown in Figure 2.3. Pump pulses of about 100 psec in duration can be converted to dye laser pulses of several picosecond or less. The dye laser gain dynamics involved in this process are illustrated in Figure 2.4. Gain increases as the integral of the pump pulse until it exceeds the losses. The pulse within the dye laser then arrives and drives the gain below the loss line. Actual pulse-shapes are very critically dependent upon the round-trip time (hence arrival time) of the dye laser pulse (Ausschnitt and Jain, 1978). Sub-picosecond pulse durations are possible but very difficult to obtain in practice. Recent streak camera results (Shapiro et al., 1981) have shown pulse break-up and satellite pulse formation when

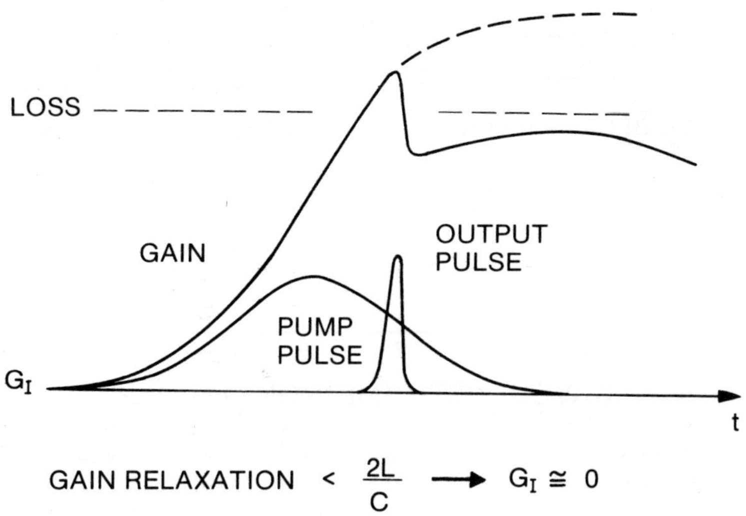

FIGURE 2.4 Gain dynamics as a synchronously-pumped dye laser.

ultrashort pulse operation is attempted. Satellite pulses form as the gain rises above the loss line for a second time. Some improvement in this regard may be achieved using a very small radius end mirror so that the gain jet is within several millimeters of the end of the resonator (Heritage et al., 1981). Then the dye laser pulse makes two passes through the gain in rapid succession.

Figure 2.5 indicates the broad spectral coverage provided by a relatively small number of dyes. Of increasing importance is the 800-900 nm region (Kuhl et al., 1977) of interest for time-resolved spectroscopy of GaAs and GaAℓAs materials and structures. Dyes in this range have tended to be unstable and to require unpleasant solvents. Very recently, stable and efficient operation from 760 nm to 900 nm has been achieved with the single dye Oxazine 750 dissolved with propylene carbonate in ethylene glycol (Aumiller, 1982; Eilenberger et al.,

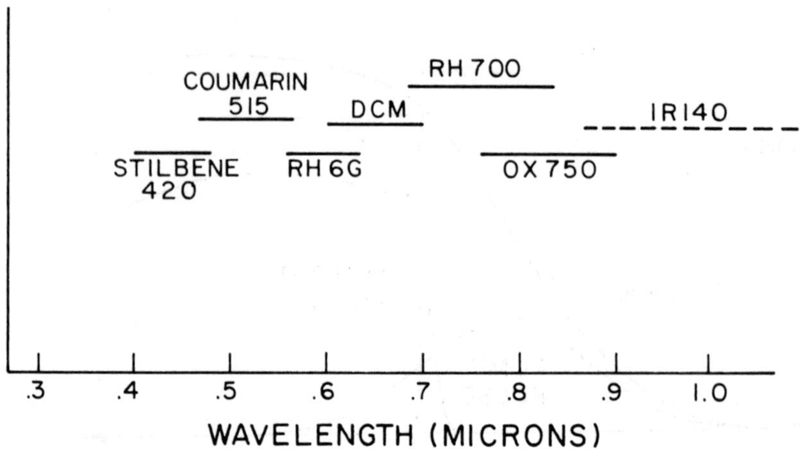

FIGURE 2.5 Efficient dyes for wavelengths spanning the visible and near infrared.

1982). Bandwidth limited pulses of about 5 psec in duration have been obtained over most of the range.

The ring geometry has been utilized by several groups for synchronous modelocking (Heritage and Isaacs, 1981; Halbout et al., 1981; Eilenberger et al., 1982), but the advantages of colliding pulse interaction here are still less clear than they are for passive modelocking.

Synchronous pumping also opens the possibility of operating two modelocked dye lasers simultaneously for experiments requiring pulses at two wavelengths. The two lasers can be operated in parallel (Heritage and Jain, 1978) pumped by the same argon laser, or in tandem (Jain and Heritage, 1975), where the first dye laser pumps the second. Most recently, a system providing two separately tunable, highly stable, cavity-dumped trains of 8 psec duration and low jitter has been developed (Heritage et al., 1982). Such a system will find many important applications in nonlinear, particularly Raman, spectroscopy (Heritage, 1980).

2.4 Synchronously-Pumped Color Center Laser

Although not yet in such wide usage as dye lasers, color center lasers offer many similar qualities, i.e. tunability and efficiency. Most importantly, they can provide cw picosecond pulse trains at wavelengths unreachable with dyes. An F_2^+ color center pumped by Nd:YAG and operating in the range 1.25 - 1.45 μm (Mollenauer and Bloom, 1979) made possible the first demonstration of long distance transmission of picosecond pulses in optical fibers (Bloom et al., 1979). Very recently, discovery of a new $T\ell^0(1)$ color center in $KC\ell$ (Gellerman et al., 1981) has made possible (also with Nd:YAG pumping) the development of a picosecond source for the range 1.4 μm to 1.6 μm (Mollenauer et al., 1982). Both lasers are valuable diagnostic tools for this long wavelength regime. The new $T\ell^0(1)$ center in particular is stable at room temperature and, under laser excitation, free from orientational bleaching.

From a modelocking point of view the $T\ell^0(1)$ centers in $KC\ell$ have interesting properties. In contrast to dyes they have a relatively low gain cross-section and a long (1600 nsec) radiative decay time. As a result, the shortest pulses are produced with synchronous pumping as far as possible above threshold in a high Q resonator (Mollenauer et al., 1982). The gain dynamics of such a system are illustrated in Figure 2.6. The long gain lifetime implies little relaxation in a round-trip time. Thus the gain prior to each pumping pulse is essentially clamped to its value following the previous lasing pulse. This has the important consequence of both stabilizing pulse energy and eliminating satellite pulses (Mollenauer et al., 1982; Haus, 1982b).

2.5 Passively Modelocked CW Dye Laser

The first sub-picosecond pulses were generated by the passively modelocked cw dye laser. A version of this laser that has

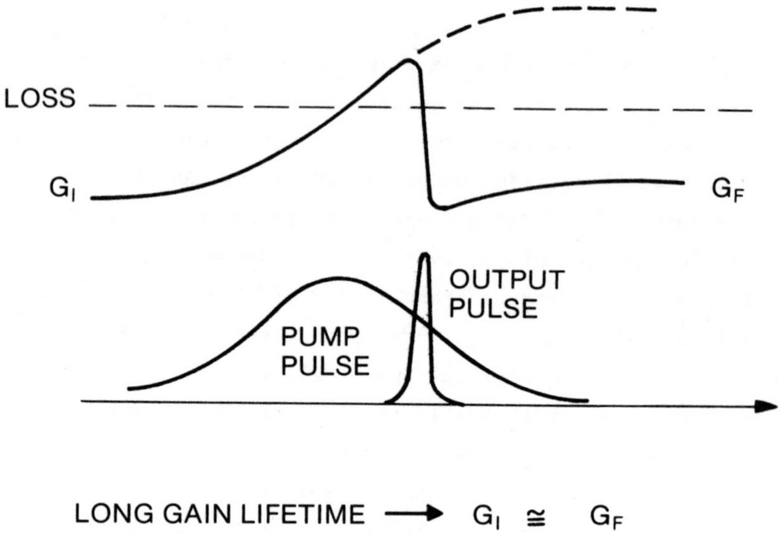

FIGURE 2.6 Gain dynamics in a synchronously pumped color center laser with a long gain lifetime.

been used successfully in a variety of picosecond studies is shown in Figure 2.7. The gain medium, rhodamine 6G in a free-

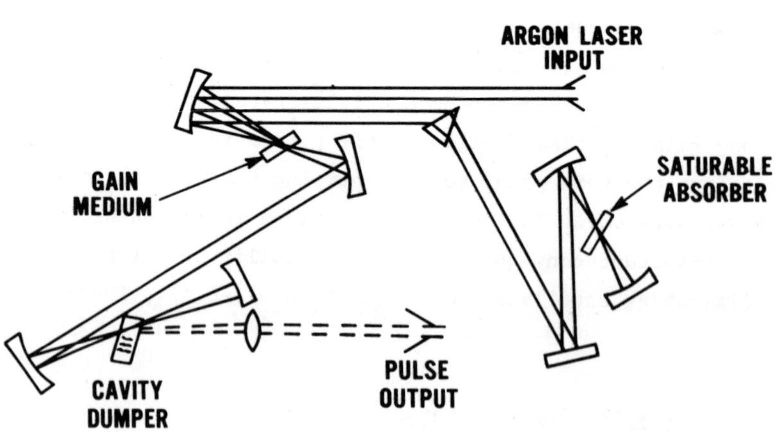

FIGURE 2.7. Cavity-dumped, passively modelocked cw dye laser.

flowing stream of ethylene glycol, is located at a focal point approximately in the center of the resonator. Near one end of the resonator is a second, free-flowing stream of ethylene glycol containing the saturable absorber dyes for modelocking. The beam waist in this stream is about a factor of two smaller than that in the gain medium. Sub-picosecond pulses can be generated with only the dye DODCI in the modelocking stream, but in this linear laser, addition of a second dye, malachite green, improves reproducibility and stability (Ippen and Shank, 1975a). The recovery time of DODCI in this system has been measured to be 1.2 nsec (Shank and Ippen, 1975), confirming its behavior as a "slow" absorber. Malachite green provides some absorption recovery on a picosecond time scale (Ippen et al., 1976) discriminating against satellites and facilitating sub-picosecond pulse operation in a stable regime well above threshold.

At the other end of the resonator is a commercially available cavity dumper that can dump single pulses from the laser at a rate of more than 10^5 pps. This arrangement is often preferable to taking the pulse output through a partially transmitting mirror. The peak power is at least an order of magnitude greater, and the lower repetition rate can be adjusted to allow for complete sample recovery (or replacement) between pulses.

The output pulses of the system shown in Figure 2.7 have been well characterized. Sub-picosecond pulses are generated near 615 nm. Energy in each pulse is about 5×10^{-9} J and the peak power exceeds 5 kW. In a focused beam, energy densities of several millijoules per square centimeter are easily achieved. When the highest temporal resolution is required, pulses can be compressed and filtered (Ippen and Shank, 1975a) to produce bandwidth-limited pulses with a duration of 0.3 psec.

The shortest of all pulses are produced by the colliding-pulse-modelocked (CPM) ring laser (Fork et al., 1981). A system in our laboratory similar to that of Fork et al., is illustrated in Figure 2.8. The prism and cavity dumper have been removed to

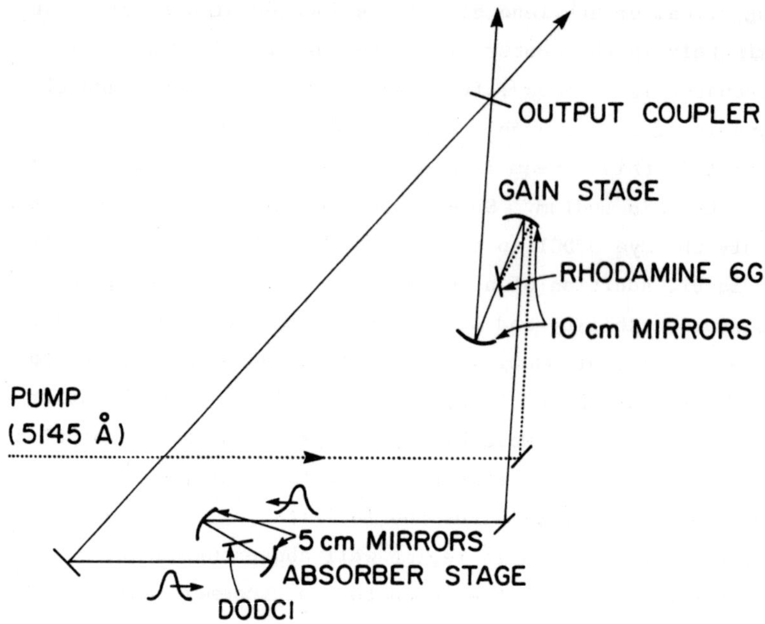

FIGURE 2.8 Femtosecond pulse generating CPM ring laser.

reduce dispersion. Since pulse collision in the saturable absorber is energetically favored (Fork et al., 1981), the RH6G gain stream is placed one-fourth of the way around the resonator so that oppositely traveling pulses are equally spaced in arrival time at the gain. The absorber stream, requiring now only DODCI, is made as thin as possible ($\sim$ 20 μm) to maximize the coherent coupling enhancement for ultrashort pulses. With a selected set of mirrors the CPM laser in our laboratory generates pulses approximately 80 fsec in duration. Similar, and even somewhat shorter, pulsewidths have been reported by Shank et al. (1982) and Halbout and Tang (1982). Output pulse duration can be greatly influenced by mirror selection which can present frustrations but may also provide opportunity for still shorter pulses.

Perhaps just as impressive as the reduction in pulsewidth achieve by CPM is the improved amplitude stability. This translates directly into improved sensitivity and signal-to-noise. Recent sampling experiments performed with this laser have demonstrated sensitivity to modulation depths of 10^{-7} (Valdmanis et al., 1982).

2.6 High Power, Sub-Picosecond Amplification

Although high repetition rate and pulse reproducibility make it possible to perform a great variety of picosecond experiments with continuously operated lasers, there are several important reasons for pushing toward higher power. Some time-resolved experiments (especially in low density systems) simply require more pulse energy. High power facilitates the conversion of pulse energy to other wavelengths for more experimental versatility. Finally, well-controlled sub-picosecond pulses make possible some very high power experiments not amenable to longer pulse durations.

An amplifier system used in our laboratory in conjunction with sub-picosecond pulses from a CPM ring laser is shown in Figure 2.9 (Migus et al., 1982; Shank et al., 1982). A frequency-doubled, Q-switched Nd:YAG is fired at optimum time for the

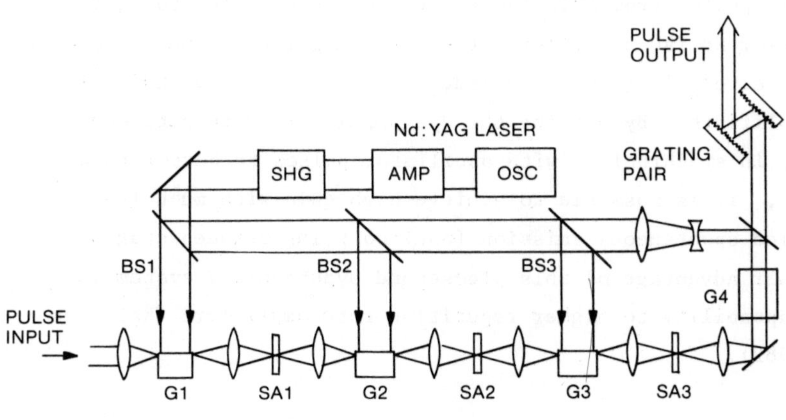

FIGURE 2.9 Four stage dye amplifier for femtosecond pulses.

selection of a single pulse from the CW dye laser. Successively larger fractions of the resulting high power pulse at 5300 Å are split off to pump four dye amplifiers. The pumping duration is about 10 nsec, synchronization with the sub-picosecond pulse is within 1 nsec, and the whole system can be operated at a rate of 10 pps.

Considerable care must be taken to prevent excessive losses due to amplified spontaneous emission (Ganiel et al., 1975). Geometrical design can be optimized to avoid self-dumping of a single stage (Wallenstein and Hansch, 1975). Strongly absorbing, malachite green, absorber streams are used to isolate stages from each other and to reshape pulses after amplification. A grating pair (see Chapter 1) is used at the output to compress the pulses by compensating for chirp introduced in the amplifiers. An optimized system such as this can produce output pulses of 80 fsec in duration and 1.2 GW peak power (Fork et al., 1982a,b).

An alternative amplification scheme for femtosecond pulses has recently been demonstrated (Kafka et al., 1982; Sizer et al., 1981). Ultrashort pulses are generated by synchronously pumping a RH6G-DQOCI (saturable absorber) dye laser with frequency-doubled pulses from a cw modelocked Nd:YAG laser. Unconverted 1.06 μm pulses from this laser are used in parallel to seed a high power Nd:YAG amplifier. The resulting intense pulses are then selected, frequency doubled, and used to pump a chain of dye amplifiers. By pumping the dye amplifiers with picosecond pulses, in synchronism with oscillator pulses to better than 50 psec, it is possible to achieve high gain with much less amplified spontaneous emission (ASE) coupling between stages. A further advantage of this picosecond synchronized system is its adaptability to higher repetition rate amplifiers (Kafka et al., 1982).

2.7 Continuum Generation

For spectroscopic purposes, high power pulses can be used to generate other wavelengths by stimulated Raman scattering, parametric conversion, or continuum generation. The latter technique (Alfano and Shapiro, 1970, 1971) is perhaps the most dramatic. By focusing amplified pulses into a cell containing H_2O, D_2O, or $CC\ell_4$ we have generated subpicosecond (continuum) pulses with a wavelength content extending from the near ultra-violet ($\sim$ 300 nm) to the infrared ($\sim$ 1.6 μm). Measurements on selected portions of this continuum show sub-picosecond pulse durations. With all reflective optics and generation in a thin (300 μm) jet of ethylene glycol, continuum pulses as short as 65 fsec and with negligible chirp over a broad range have been demonstrated (Fork et al., 1982b). This continuum, with an optical multichannel analyzer to obtain high resolution time spectra (Shank et al., 1978a), is now being applied to a variety of molecular and solid state studies.

REFERENCES

Alfano, R.R., and S.L. Shapiro, 1970, Phys. Rev. Lett. 24, 584.

Alfano, R.R., and S.L. Shapiro, 1971, Chem. Phys. Lett. 3, 407.

Attwood, D.T., L.W. Coleman, M.J. Boyle, J.T. Larsen, D.W. Phillion, and K.R. Manes, 1977, Phys. Rev. Lett. 38, 282.

Aumiller, G.D., 1982, Opt. Commun. 41, 115.

Ausschnitt, C.P., and R.K. Jain, 1978, Appl. Phys. Lett. 32, 727.

Bloom, D.M., L.F. Mollenauer, C. Lin, D.W. Taylor, and A.M. DelGaudio, 1979, Opt. Lett., 297.

Bradley, D.J., New, G.H.C., and S.J. Caughey, 1970, Opt. Commun. 2, 41.

Campillo, A.J., R.C. Hyer, and S.L. Shapiro, 1979, Opt. Lett. 4, 325.

DeMaria, A.J., D.A. Stetser, and H. Heynau, 1966, Appl. Phys. Lett. 8, 22.

Duguay, M.A., J.W. Hansen, and S.L. Shapiro, 1970, IEEE J. Quant. Electron.

Eckhardt, R.C., C.H. Lee, and J.N. Bradford, 1974, Opto. Electron. 6, 67.

Eilenberger, D.J., E. Isaacs, and G.D. Aumiller, 1982, to be published.

Fendt, A., W. Kranitsky, A. Laubereau, and W. Kaiser, 1979, Opt. Commun. 24, 237.

Fork, R.L., B.I. Greene, and C.V. Shank, 1981, Appl. Phys. Lett. 38, 671.

Fork, R.L., C.V. Shank, and R.T. Yen, 1982a, Appl. Phys. Lett. 41, 273.

Fork, R.L., C.V. Shank, R.T. Yen, and C. Hirlimann, 1982b, *Picosecond Phenomena III* (Springer Verlag).

Ganiel, U., A. Hardy, G. Neumann, and D. Treves, 1975, IEEE J. Quant. Electron. QE-11, 881.

Gellerman, W., F. Luty, and C.R. Pollack, 1981, Opt. Commun. 39, 391.

Glenn, W.H., 1975, IEEE J. Quant. Electron. QE-11, 8.

Halbout, J.M., and C.L. Tang, 1982, Appl. Phys. Lett. 40, 765.

Halbout, J.M., A. Olsson, and C.L. Tang, 1981, Appl. Phys. Lett. 39, 463.

Haus, H.A., 1982b, unpublished.

Heritage, J.P., 1980, *Picosecond Phenomena II* (Springer Verlag), 343.

Heritage, J.P., and R. Jain, 1978, Appl. Phys. Lett. 32, 101.

Heritage, J.P., and E.D. Isaacs, 1981, Technical Digest CLEO.

Heritage, J.P., E. D. Isaacs, and E.P. Ippen, 1981, unpublished.

Heritage, J.P., D.S. Chemla, and P.F. Liao, 1982, *Picosecond Phenomena III* (Springer Verlag).

Ippen, E.P., and C.V. Shank, 1975a, Appl. Phys. Lett. 27, 488.

Ippen, E.P., C.V. Shank, and A. Dienes, 1972, Appl. Phys. Lett. 21, 348.

Ippen, E.P., C.V. Shank, and A. Bergman, 1976, Chem. Phys. Lett. 38, 611.

Jain, R., and J.P. Heritage, 1975, Appl. Phys. Lett. 32, 41.

Kafka, J.D., T. Sizer, C.W. Gabel, and G. Mourou, 1982, *Picosecond Phenomena III* (Springer Verlag).

Kaiser, W., and A. Laubereau, 1977, *Nonlinear Optics*, edited by Harper and Wherrett (Academic Press), 257.

Kaiser, W., A. Fendt, W. Kranitzky, and A. Laubereau, 1980, Phil. Trans. R. Soc. (London) A298, 267.

Korobkin, V.A., A.A. Malyutin, and A.M. Prokhorov, 1970, JETP Lett. 12, 150.

Kryukov, P.G., and V.S. Letokhov, 1972, IEEE J. Quant. Electron. QE-8, 766.

Kuhl, J., R. Lambrich, and D. von der Linde, 1977, Appl. Phys. Lett. 31, 657.

Laubereau, A., L. Greiter, and W. Kaiser, 1974, Appl. Phys. Lett., 25, 87.

Migus, A., C.V. Shank, E.P. Ippen, and R.L. Fork, 1982, IEEE J. Quant. Electron. QE-18, 101.

Mollenauer, L.F., and D.M. Bloom, 1979, Opt. Lett. 4, 247.

Mollenauer, L.F., N.D. Vieiro, and L. Szeto, 1982, to be published.

Reintjes, J., C.Y. She, R.C. Eckhardt, N.E. Karangelen, R. Andrews, and R. Elton, 1977, Appl. Phys. Lett. 30, 489.

Seilmeier, A., K. Spanner, A. Laubereau, and W. Kaiser, 1978, Opt. Commun. 24, 237.

Shank, C.V., and E.P. Ippen, 1975, Appl. Phys. Lett. 26, 62.

Shank, C.V., R.L. Fork, R.F. Leheny, and J. Shah, 1978, Phys. Rev. Lett. 42, 112.

Shank, C.V., R.L. Fork, and R.T. Yen, 1982, *Picosecond Phenomena III* (Springer Verlag).

Shapiro, S.L., 1977, *Ultrashort Light Pulses* (Springer Verlag), chapters 2 and 3.

Shapiro, S.L., R.R. Cavanaugh, and J.C. Stephenson, 1981, Opt. Lett. 6, 471.

Siegman, A.E., 1973, IEEE J. Quant. Electron. QE-9, 247.

Sizer, T., J.D. Kafka, A. Krisiloff, and G. Mourou, 1981, Opt. Commun. 39, 259.

Vanherzeele, H. and A.E. Siegman, 1982, Technical Digest CLEO.

Von der Linde, D., 1972, IEEE J. Quant. Electron. QE-8, 328.

Von der Linde, D., O. Bernecker, and A. Laubereau, 1970, Opt. Commun. 2, 215.

Wallenstein, R. and T.W. Hansch, 1975, Opt. Commun. 14, 353.

Zinth, W., A. Laubereau and W. Kaiser, 1977, Opt. Commun. 22, 161.

3. TECHNIQUES OF PICOSECOND MEASUREMENT

3.1 Introduction

With the advent of picosecond optical pulses it became apparent that conventional electronics could not provide sufficient time resolution either for measuring pulses or for studying the dynamics of their interactions with matter. The rapid development of pulse measurement techniques (Ippen and Shank, 1977) that followed relied mostly on the use of nonlinear optical interactions. In a nonlinear material one picosecond pulse can be used to probe either itself or another pulse, and the resulting measurement is a function of autocorrelation or cross-correlation. Likewise, in studies of picosecond phenomena one pulse can be used to 'probe' the state of the medium following excitation by a 'pump' pulse or to 'gate' the response of a detection scheme. Picosecond streak cameras (Bradley, 1977) have also provided valuable information on pulse characteristics and are currently finding increasing application to studies of time-resolved luminescence.

3.2 Pulse Diagnostics

Autocorrelation measurements by second harmonic generation (SHG) (Weber, 1967; Maier et al., 1967; Armstrong, 1967) and two-photon fluorescence (TPF) (Giordmaine et al., 1967) are by far the most common means for evaluating ultrashort pulses. In the case of TPF there is a nonlinearly enhanced absorption (followed by fluorescence) and in SHG a coherent (usually phasematched) generation of a signal proportional to

$$\langle |\{E(t) + \sqrt{b}\, E(t+\tau)\, e^{j\omega\tau}\}^2|^2 \rangle \qquad (3.1)$$

where τ is the delay between the two pulses and 'b' is the ratio of their intensities. The measured signal, with τ averaged over an optical period, is

$$<E^4(t) + b^2E^4(t + \tau) + 4b\ E^2(t)\ E^2(t + \tau)>$$

$$= (1 + b^2)<I^2(t)> + 4b<I(t)\ I(t + \tau)>. \quad (3.2)$$

Normalizing to the background, we have for the measurement

$$f(\tau) = 1 + \frac{4b}{(1 + b^2)} G^2(\tau) \quad (3.3)$$

where $G^2(\tau)$ is the second order autocorrelation function of the intensity:

$$G^2(\tau) = \frac{<I(t)\ I(t + \tau)>}{<I^2(t)>} . \quad (3.4)$$

Several comments can be made on the interpretation of experimental results. $G^2(\tau)$ is by definition equal to unity at $\tau = 0$ so that well isolated pulses with $b = 1$ give a contrast ration $f(0)$: $f(\infty)$ of 3:1. Accurate measurement of this contrast ratio is important since even a purely random signal has an autocorrelation peak at $\tau = 0$ and a variety of statistical behavior is possible (Rowe and Li, 1970). $G^2(\tau)$ is also by definition symmetric in τ. This is a fundamental limitation on the information one can derive about the pulse shape $I(t)$. Higher order correlation measurements and other probing methods are necessary to determine pulse asymmetry (Auston, 1971; von der Linde et al., 1971; Ippen and Shank, 1975a). Nevertheless, if a proper measure of $G^2(\tau)$ is obtained, one can estimate the pulse duration to better than a factor of two.

Idealized traces for $f(\tau)$ and $G^2(\tau)$ for different types of optical signals are shown in Figure 3.1. To derive a pulse duration from $G^2(\tau)$ it is necessary to assume a pulseshape. The relationship between autocorrelation width and

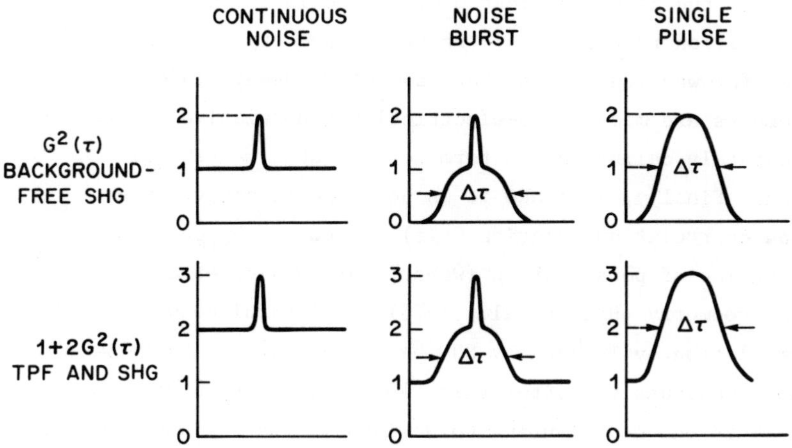

FIGURE 3.1 Autocorrelation traces for transform limited and noisy signals.

pulsewidth are shown in Figure 3.2. Also shown in Figure 3.2 is the Fourier-transform limited spectral width for each of the pulses. All measurements should be consistent if the pulse is to be called transform limited.

$I(t)$	$\Delta\tau/\Delta t$	$\Delta t \Delta \nu$
$1 \ (0 \le t \le \Delta t)$	1	.886
$\exp\left\{-\frac{(4\ln 2)t^2}{\Delta t^2}\right\}$	$\sqrt{2}$	.441
$\mathrm{sech}^2\left\{\frac{1.76\,t}{\Delta t}\right\}$	1.55	.315
$\exp\left\{-\frac{(\ln 2)t}{\Delta t}\right\}$ $(t \ge 0)$	2	.11

FIGURE 3.2 Autocorrelation widths, pulsewidths, and spectral widths for different pulse shapes.

SHG is the technique most widely used in conjunction with high repetition rate sources. Phase-matched crystals are easily obtained for wavelengths as short as 400nm (Dewey, 1976). Measurements are made photo-electrically and relative delays are accurately calibrated in terms of variable mechanical position. Finally, with SHG it is possible to measure the pulse autocorrelation function $G^2(\tau)$ without background by proper choice of polarization (Weber, 1967; Maier et al., 1967) or geometry (Haus et al., 1975). A careful comparison of autocorrelation by TPF and by SHG have shown agreement between the two techniques to better than 10^{-13} sec (Ippen et al., 1977). When one is measuring pulsewidth in the femtosecond regime, however, it is important to consider carefully all bandwidth limiting, pulse dispersing, and geometric overlap effects.

A particular experimental arrangement that has been used successfully with high repetition rate systems is shown in Fig. 3.3. The picosecond pulse train is divided into two beams which traverse different path lengths in a modified interferometer. A stepper-motor-driven translation stage provides a variable path length in one of the arms and allows a digitally controllable scan of the desired delay. The two beams, with relative time delay, leave the interferometer parallel but noncollinear. A simple lens focuses both beams to the same spot in an angle-phase-matched non-linear crystal, usually KDP (but $LiIO_3$ for better sensitivity with diode lasers). SHG, proportional only to the product of the two beams, is detected at an angle bisecting that of the individual beams. The signal as a function of relative delay is stored in a multichannel analyzer and averaged by repeated scanning over the desired range. Good autocorrelation traces are obtained in a matter of minutes; the same system provides averaging over hours for applications where the signal is small. A pulse measurement made by this background-free autocorrelation system using SHG is shown in Figure 3.4. The pulses, from a

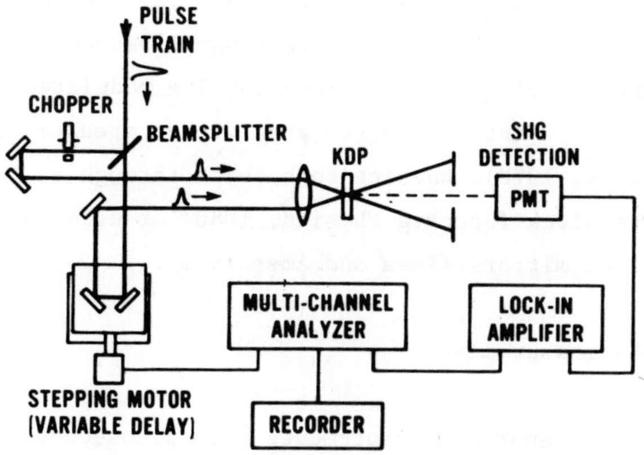

FIGURE 3.3 Experimental arrangement for accurate sub-picosecond pulse autocorrelation measurements.

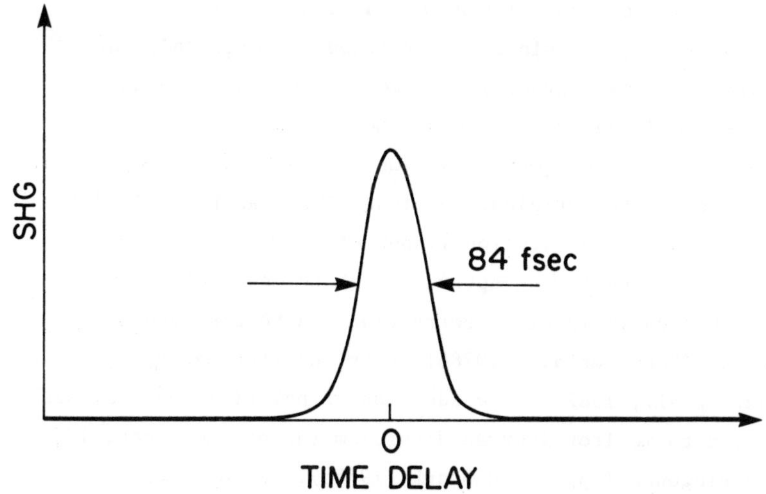

FIGURE 3.4 Autocorrelation measurement of a femtosecond pulse.

CPM cw dye laser, have a duration (FWHM) of about 80 fsec.

For real-time pulse monitoring and ease in laser alignment, several schemes now make possible direct oscilloscope display of the autocorrelation function. They involve modulating the length of one of the interferometer arms with a speaker coil (Fork and Beisser, 1978), passing both beams through a spinning quartz block (Spectra Physics, 1980) or using a pair of spinning parallel mirrors (Yasa and Amer 1981).

3.3 Pump-probe measurements

In the first reported measurements with picosecond optical pulses (Shelton and Armstrong, 1967) a beam splitter was used to pick off a small portion of the pulse train and a variable optical delay path was introduced between the two beams. The main beam was used to excite (pump) a dye sample and the weak (probe) beam was used to monitor the recovery of dye transmission as a function of delay. This pump-probe method has since been refined and extended, but it remains the most versatile and most widely used technique for pico-second and sub-picosecond measurement.

Pump probe measurements can monitor saturation, depletion, and recovery of the original species. They can be applied to studies of induced absorption (Ippen et al., 1976) due to excited states, energy acceptors, or photo-chemical products. Induced changes in index of refraction can be observed by reflection (Shank et al., 1978b) or by interferometry (Halbout et al., 1982). The addition of polarization analysis allows one to monitor induced dichroism (Shank and Ippen, 1975) or birefringence (Ippen and Shank, 1975b). With slight modification of the system shown in Figure 3.3, pump-probe experiments are performed. The sample under investigation is placed at the common focus of the pump and probe beams and the detector is moved to monitor the transmission of the probe beam through the sample. No fast detector or high speed electronics is required. All timing accuracy is obtained by control of the optical delay. Detection of signal on the probe beam is at the modulation frequency imposed upon the pump. Integration time on the lock-in amplifier is adjusted to be compatible with the scanning speed. Multiple scanning and digital averaging, rather than a long time-constant on the lock-in is often preferable because it reduces the chance of signal distortion due to drift in the system.

Consider the ultimate sensitivity of such a system. The average current in the detector is

$$\langle I \rangle = e \frac{\eta P}{\hbar\omega_o} \tag{3.5}$$

where η is the quantum efficiency, e is the electronic change, $\hbar\omega_o$ is the photon energy and P is the average optical power in the probe beam. The shot noise power due to this average current is

$$\langle I_{SN}^2 \rangle = 2e \langle I \rangle \, \Delta f \tag{3.6}$$

where Δf is the bandwidth of the detection system. For a modulation depth m, the signal power one is trying to detect is $m^2 \langle I \rangle^2$, and the signal to-noise ratio in this shot-noise-limit is

$$S/N = m^2 \left(\frac{\eta}{2\hbar\omega_o} \right) \frac{P}{\Delta f} \, . \tag{3.7}$$

As an example, for $\eta \cong .9$, an average power at 600 nm of 10mW, and a bandwidth of 1Hz, a modulation depth of about 3×10^{-8} can be detected with an S/N = 10.

It is interesting to note that the pulse stability of picosecond lasers is reaching the point where detection near this shot-noise-limit is conceivable. Several years ago, fluctuations in the laser output limited detectibility to the range

10^{-3} - 10^{-4}. The stable CPM ring laser and synchronously pumped dye lasers, modulated at high (MHz) frequencies to avoid jet stream and other acoustic noise, have improved sensitivity by factors of more than a thousand (Heritage et al., 1982; Valdmanis et al., 1982).

Accurate interpretation of pump-probe measurements near t = 0 also requires that one analyze the effects of coherent pulse interaction (Ippen and Shank, 1977). This can give rise to additional, apparently fast, contributions to the signal that have little dependence upon the response of the material under investigation. Without coherent coupling (or interference) between the pump and probe pulse, the expected average change in probe beam intensity due to the pump would be

$$\gamma(\tau) = \int_{-\infty}^{\infty} A_{\gamma}(\tau - \tau') \int_{-\infty}^{\infty} I(t)\ I(t + \tau')\,dt \tag{3.8}$$

which is just the autocorrelation function (instantaneous response) convolved with the impulse response of the material change that couples the two beams. Equation 3.8 ignores the spatial and polarization interference effects that are present when the two pulses overlap and are coherent. The additional "coherent artifact" (Ippen and Shank, 1977) due to this interference has been observed in a variety of experiments (Shank and Auston, 1975; Ippen et al., 1976; Shank and Ippen, 1975) and may be written in the form

$$\beta(\tau) = \int_{-\infty}^{\infty} E^*(t)\ E(t + \tau) \int_{-\infty}^{\infty} E^*(t')\ E(t' + \tau)\ A_{\beta}(t - t')\,dt'\,dt. \tag{3.9}$$

One may think of this term as being the scattering of the pump beam into the probe beam by the periodic nonlinear polarization they produce in the medium. It is precisely the term we discussed in Chapter 1 for the case of CPM modelocking. It is important to point out that A_{γ} and A_{β} are not necessarily equal if the pump and probe beams are of different polarization.

For example, reorientation (or loss of polarization memory within the medium) on a time scale short compared to the pulse will greatly reduce this term (von Jena and Lessing, 1979; Vardeny and Tauc, 1981). Studies of the dynamics of such gratings, induced by two pulses and probed by a third pulse (Phillion et al., 1975), add yet another important dimension to pump-probe spectroscopy.

Often it is desired to measure an entire absorption spectrum in a single shot using a picosecond interrogating pulse with a broad optical bandwidth. Alfano and Shapiro, 1970, first suggested that a picosecond continuum, generated when an intense picosecond laser pulse is focused into a non-linear medium, could be used for monitoring absorption spectra of transient species on a picosecond time scale. Malley, 1974, described the use of a continuum in a configuration that allows one to take both time and wavelength resolved data with a single pulse system. Now with amplified femtosecond dye laser pulses at 10 Hz and the femtosecond continuum they can generate (Fork et al., 1982), time and wavelength resolved spectra can be averaged extensively and be used to expose previously unresolvable physical phenomena (Shank et al., 1982).

3.4 Optical Gating

A scheme of extending the pump-probe technique to wavelengths at which synchronized probe pulses are not available is shown in Figure 3.5 (Wiesenfeld and Ippen, 1979). Pulses from the subpicosecond laser are divided into two beams. One beam, the 'pump', traverses a variable path delay and is aligned parallel, with the output beam of a cw (or cavity dumped) tunable laser, the 'probe'. This probe beam experiences changes in the sample due to the ultrashort pumping pulse. The transient change is then detected by using the second beam from the subpicosecond source to 'gate' a subpicosecond segment

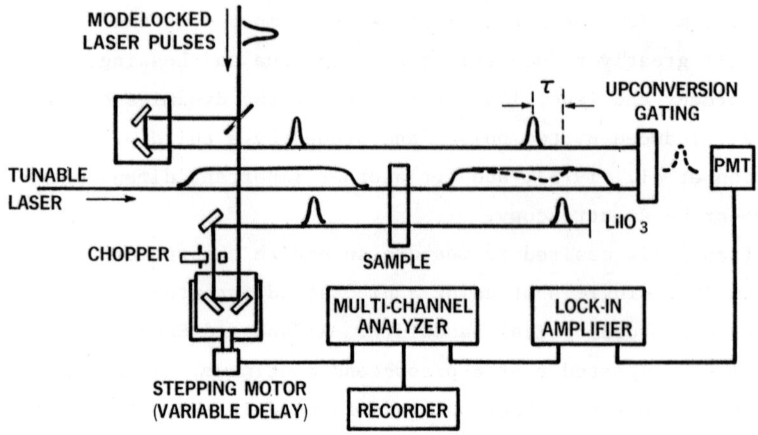

FIGURE 3.5 Experimental arrangement for tunable probe grating with subpicosecond resolution.

of the probe by sum-frequency upconversion. With synchronous detection, the output is only proportional to the change induced by the chopped pumping pulse beam. Signals are averaged as the delay is automatically and repetitively scanned by a stepping motor.

Tunable probe gating preserves the time resolution and sensitivity of the modelocked cw source. Signal-to-noise ratios in experiments performed with the tunable probe are often as good as those obtained in experiments with 615 nm pulses alone, because the cw laser introduces negligible amplitude noise and the sum-frequency generation is relatively efficient. With $LiIO_3$ and our 615 nm gating pulses, we can upconvert probe wavelengths (λ > 560 nm) with an efficiency of about 10%. For shorter wavelengths it becomes necessary to use a crystal like KDP for which our conversion efficiency is about 0.3%. A particularly important area of application for this technique is in the infrared where even subpicosecond continuum probing cannot presently reach (Wiesenfeld et al.,

1981).

Frequency up-conversion by phase-matched sum-frequency generation can be used to detect and time-resolve sample fluorescence (Duguay and Hansen, 1968). This type of gate has been used successfully to study temporal dynamics of synchronously pumped lasers (Mahr and Hirsch, 1975) and relaxation processes in large molecules (Hallidy and Topp, 1977). Advantages of up-conversion are that it is fast (i.e. it is not limited by any response time of the medium) and that it is background-free (i.e. there is no signal leakage to the detector in the absence of a gating pulse). The latter fact becomes paramount if one tries to sample a subpicosecond interval within a nanosecond-long emission period. A disadvantage is that, because phase matching is both angle and frequency selective, only a very small fraction of the emission is up-converted. For sub-picosecond studies this will probably require photon-counting techniques.

A gating scheme that offers broadband frequency transmission and larger angular acceptance is the optical Kerr shutter (Duguay and Hansen, 1969). Basic components for such a shutter are shown in Figure 3.6. An arrangement used for

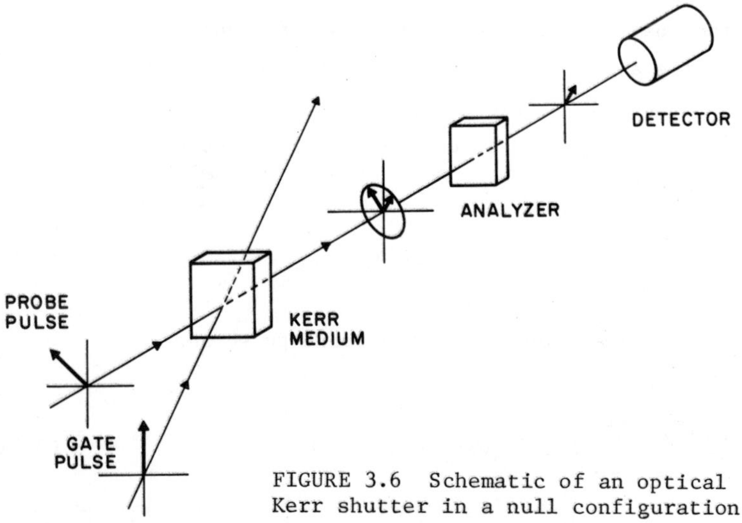

FIGURE 3.6 Schematic of an optical Kerr shutter in a null configuration.

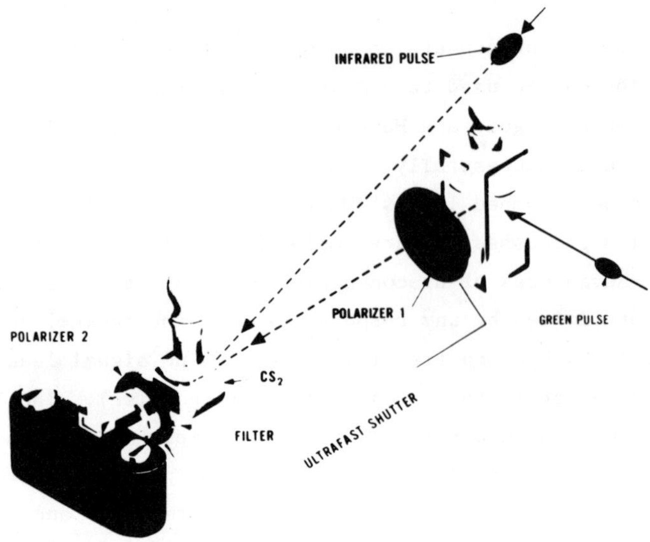

FIGURE 3.7 Experimental arrangement for the photography of light in flight (Duguay and Hansen, 1971).

photography of light in flight (Duguay and Hansen, 1971) and for gated picture ranging (Duguay and Savage, 1973) is shown in Figure 3.7.

If the analyzer in Figure 3.6 is set for null transmission in the absence of a pump pulse (a minimum transmission of about 10^{-5} between good, crossed polarizers determines the background level) the time integrated transmission is given by

$$\varepsilon(\tau) = \int_{-\infty}^{\infty} I_p(t + \tau) \sin^2\{\phi(t)/2\} \, dt \tag{3.10}$$

where $I_p(\tau)$ is the intensity of the probe pulse and

$$\phi(t) = \frac{2\pi L}{\lambda} \frac{n_{2B}}{\tau_R} \int_{-\infty}^{t} I_G(t') \exp\{-(t - t')/\tau_R\} \, dt' \tag{3.11}$$

where n_{2B} is the nonlinear birefringence coefficient, τ_R is the response time of the birefringent change involved and $I_G(t)$ is the intensity of the gating pulse.

With cw lasers it is possible to operate a Kerr gate biased for 50% transmission with a $\lambda/4$ plate (Ippen and Shank, 1975b). Phase-lock detection is used to electronically eliminate the background and the integrated detected signal becomes

$$\varepsilon(\tau) = \int_{-\infty}^{\infty} I_p(t + \tau) \sin \phi(t) \, dt. \tag{3.12}$$

Biasing into this linear regime can improve sensitivity and eliminate temporal distortions due to pulse asymmetries (Ippen and Shank, 1975b). This latter scheme has been used recently (Greene, 1982) to study femtosecond relaxation processes in CS_2 (Halbout et al., 1982).

REFERENCES

Alfano, R.R. and S.L. Shapiro, 1970, Phys. Rev. Lett. 24, 584.

Armstrong, J.A., 1967, Appl. Phys. Lett. 10, 16.

Auston, D.H., 1971, Appl. Phys. Lett. 18, 249.

Bradley, D.J., 1977, *Ultrashort Light Pulses*, Topics in Appl. Phys., edited by S.L. Shapiro (Springer Verlag), vol. 18.

Dewey, H.J., 1976, IEEE J. Quant. Electron. QE-12, 303.

Duguay, M.A., and J.W. Hansen, 1968, Appl. Phys. Lett. 13, 178.

Duguay, M.A., and J.W. Hansen, 1969, Appl. Phys. Lett. 15, 192.

Duguay, M.A., and J.W. Hansen, 1971, Amer. Scient. 59, 550.

Duguay, M.A., and A. Savage, 1973, Opt. Commun. 9, 212.

Fork, R.L. and F.A. Beisser, 1978, Appl. Opt. 17, 3534.

Fork, R.L., C.V. Shank, R.T. Yen, and C. Hirlimann, 1982b, *Picosecond Phenomena III* (Springer Verlag).

Greene, B.I., 1982, *Picosecond Phenomena III* (Springer Verlag).

Halbout, J.M., and C.L. Tang, 1982, Appl. Phys. Lett. 40, 765.

Hallidy, L.A. and M.R. Topp, 1977, Chem. Phys. Lett. 46, 8.

Haus, H.A., C.V. Shank, E.P. Ippen, 1975, Opt. Commun. 15, 29.

Heritage, J.P., D.S. Chemla, and P.F. Liao, 1982, *Picosecond Phenomena III* (Springer Verlag).

Ippen, E.P. and C.V. Shank, 1975a, Appl. Phys. Lett. 27, 488.

Ippen, E.P. and C.V. Shank, 1975b, Appl. Phys. Lett. 26, 92.

Ippen, E.P. and C.V. Shank, 1977, *Ultrashort Light Pulses*, edited by S. L. Shapiro (Springer Verlag).

Ippen, E.P., C.V. Shank, and A. Bergman, 1976, Chem. Phys. Lett. 38, 611.

Ippen, E.P., C.V. Shank, and R.L. Woerner, 1977, Chem. Phys. Lett. 46, 20.

Mahr, M., and M.D. Hirsch, 1975, Opt. Commun. 13, 96.

Maier, M., W. Kaiser, and J.A. Giordmaine, 1966, Phys. Rev. Lett. 17, 1275.

Malley, M., 1974, *Creation and Detection of the Excited State*, edited by W.R. Ware (Marcel Dekker), vol. 2.

Phillion, D.W., D.J. Kuizenga, and A.E. Siegman, 1975, Appl. Phys. Lett. 27, 85.

Shank, C.V. and D.H. Auston, 1975, Phys. Rev. Lett. 34, 479.

Shank, C.V., and E.P. Ippen, 1975, Appl. Phys. Lett. 26, 62.

Shank, C.V., R.L. Fork, and R.T. Yen, 1982b, *Picosecond Phenomena III* (Springer Verlag).

Shelton, J.W., and J.A. Armstrong, 1967, IEEE J. Quant. Electron. QE-3, 302.

Spectra Physics, 1980, Model 409 Scanning Autocorrelator.

Valdmanis, J.A., G. Mourou and C.W. Gable, 1982, *Picosecond Phenomena III* (Springer Verlag).

Vardeny, Z. and J. Tauc, 1981, Opt. Commun. 39, 396.

Von der Linde, D., A. Laubereau, and W. Kaiser, 1971, Phys. Rev. Lett. 26, 954.

Von Jena, A. and H.E. Lessing, 1979, Appl. Phys. 19, 131.

Weber, H.P., 1967, J. Appl. Phys. 38, 2231.

Wiesenfeld, J.M. and E.P. Ippen, 1979, Chem. Phys. Lett. 67, 213.

Wiesenfeld, J.M. L.F. Mollenauer, and E.P. Ippen, 1981, Phys. Rev. Lett., 47, 1968.

Yasa, Z.A., and N.M. Amer, 1981, Opt. Commun. 36, 406.

4. ULTRAFAST PROCESSES IN LIQUIDS AND SOLIDS

4.1 Introduction

Ultrashort pulses are finding increasing application in areas of physics and chemistry as well as biology. Three specific properties have made them especially useful: short duration, coherence, and high peak power. The first, has enabled scientists to discover and study ultrafast, fundamental material processes. Temporal measurements are particularly important to improving understanding of relaxation processes in molecules, fluctuations in liquids, transport and relaxation in solid state materials, and dynamics of highly excited plasmas. These are all areas in which ultrafast phenomena do not easily lend themselves to study in the frequency domain (i.e. by measurements of spectral widths and shapes). Pulse coherence adds the dimension of being able to investigate transient coherent phenomena (Hesselink and Wiersma, 1978; Olson et al., 1981) and to separate dephasing from energy relaxation (Laubereau et al., 1972). The high power property enhances a multitude of nonlinear optical effects (Auston, 1977), allows the study of dynamical processes following ultra-high excitation, and facilitates a variety of nonlinear spectroscopic techniques.

Just as the invention of new devices and techniques has opened up new areas for scientific investigation, so has an improved understanding of fast phenomena led to the development of new devices. Compact picosecond semiconductor sources have been demonstrated and optical devices for ultra-high-speed signal processing are currently the focus of considerable attention. The generation of picosecond electrical pulses (Auston, 1975) provides a means for combining optics and electronics for obtaining valuable new information about very high speed microelectronic devices.

The range of picosecond applications is too large for us to attempt a comprehensive review. We therefore limit ourselves to several specific examples, which nevertheless indicate the interdisciplinary nature of this field.

4.2 Molecular Dynamics

The study of nonradiative relaxation in organic molecules is a natural application of picosecond optical pulses. Intramolecular relaxation rates can be extremely fast in large molecules. Determining them is of theoretical scientific interest and of practical interest for picosecond lasers and diagnostic tools. The basic experimental arrangement for time-resolved, pump-probe spectroscopy is shown in Figure 4.1. Measurements should be performed repetitively, at high rate, but with sufficient time between pulse pairs to allow for complete recovery. An early application of this scheme was the evaluation of recovery dynamics in saturable absorber dyes. A measurement of the 2.1 psec response of malachite green in methanol to saturation at 6.15 nm is shown in Figure 4.2 (Ippen et al., 1976). The rapid recovery of this dye has made it an important component of sub-picosecond dye oscillators and amplifiers (see Chapter 2). Its complicated dependence upon solvent viscosity

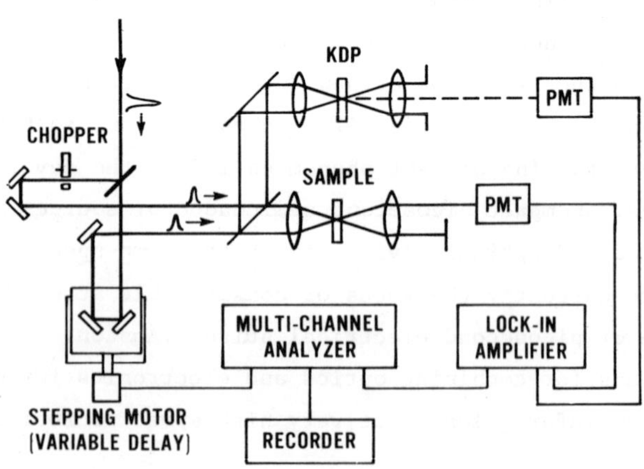

FIGURE 4.1 Experimental arrangement for time-resolved spectroscopic measurement at high repetition rate.

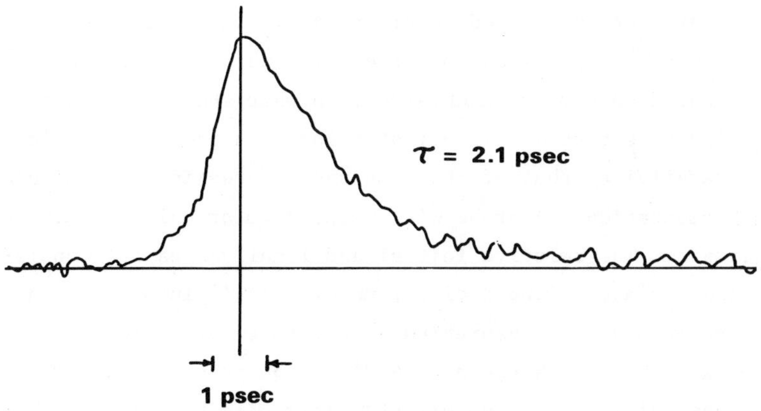

FIGURE 4.2 Picosecond absorption recovery of the dye malachite green.

and excitation wavelengths have made it (and other triphenylmethane dyes) the object of several interesting studies (Yu et al., 1977; Sundström et al., 1982).

By using different probe polarizations, it is possible with the same experimental arrangment to study molecular orientation

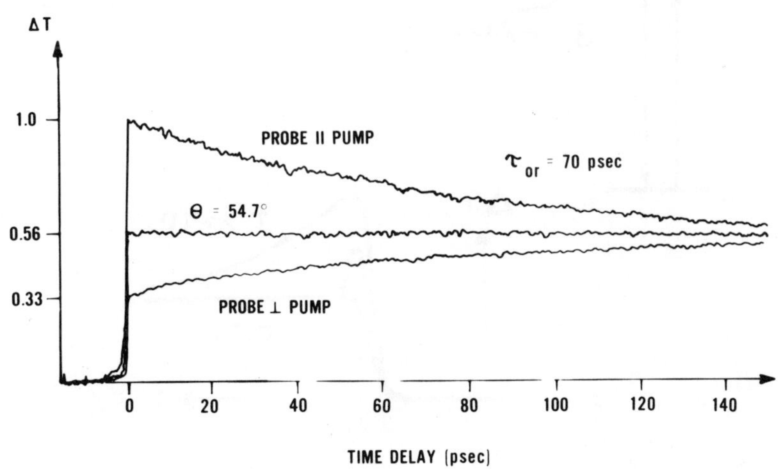

FIGURE 4.3 Orientational relaxation of Nile Blue observed with different probe polarizations.

dynamics. A set of experimental curves obtained with the laser dye Nile Blue are presented in Figure 4.3. The excited state lifetime of this dye is on the order of several nanoseconds, long compared to timescale observed, so that the predominant relaxation to Figure 4.3 is orientational. A probe with polarization parallel to that of the pump sees decreasing transmission with reorientation. A probe with perpendicular polarization sees an increase. The relative initial and final values are consistent with a classic linear dipole model. It is intuitive, and both experimental and theoretical fact, that for a probe polarization somewhere in between, 54.7° to be precise, the probe can be insensitive to these orientational effects. Alternatively, the anisotropic component can be enhanced by making the detector sensitive only to changes in polarization (Shank and Ippen, 1975; Waldeck et al., 1981).

Analysis of the response near $t = 0$ in this experiment can give an indication of molecular dynamics within the excited

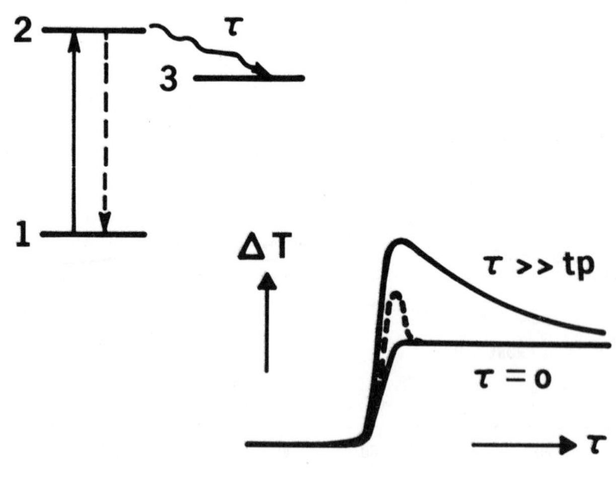

FIGURE 4.4 Saturation dynamics of a simple 3-level system.

state. Consider the dye molecule as simply a three level system like that in Figure 4.4. Excitation occurs from the S_0 ground state to some excited vibrational state of S_1 manifold. Relaxation out of this initial state occurs with a time τ. If this time behavior is easily resolved ($\tau >> t_p$ in Figure 4.4), the induced transmission is seen to relax to one-half its initial value. More importantly, if τ is shorter than the pulse duration, it is still possible to detect an initial transient and to use its amplitude to estimate τ. The amplitude of the transient, for small values relative to the final value, will simply be τ/t_p. If the experimental trace of Nile Blue, taken at a 54.7° pump-probe polarization angle to accurately establish the $\tau = 0$ level, is analyzed we can place an upper limit on τ of 5×10^{-14} sec. The accuracy of this experiment is limited here by uncertainty in the shape of the coherent artifact (see Chapter 3). It demonstrates, however, that relaxation times can be exceedingly fast and that no significant hole burning occurs even on this time scale.

Investigations of fast relaxation in dye molecules have been extended by means of tunable probe gating (Wiesenfeld and Ippen, 1979). The particular experiment we discuss here is illustrated in Figure 4.5. The molecule Nile Blue A perchlorate is excited from S_0 to S_1 by a pulse at 615 nm, near the peak of the ground state (S_0) absorption. In the excited state, then, the molecule initially has an excess vibrational energy of about 800 cm^{-1}. The object of the experiment is to study the emission spectrum ($S_1 \rightarrow S_0$) as the molecule relaxes to the lowest vibrational levels of S_1. To do this, we monitor the gain spectrum with the tunable probe laser. A typical experimental result is shown in Figure 4.6.

Shown on the same graph are an experimentally determined instantaneous response and a response calculated for a 1 psec exponential rise. The gain curve for Nile Blue follows the instantaneous system response within the 0.2 psec accuracy of the experiment. Probing at other wavelengths confirmed that the

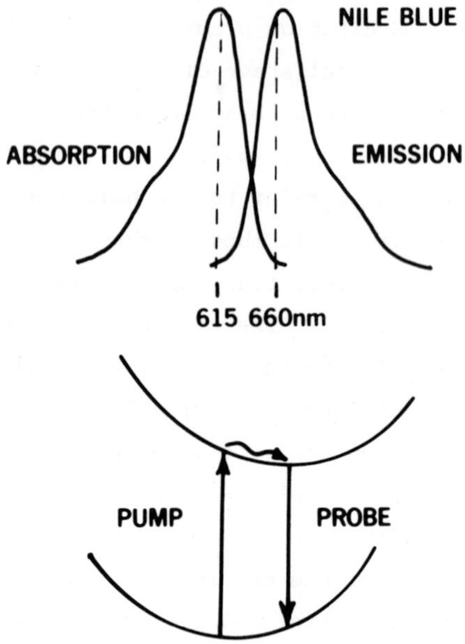

FIGURE 4.5 Spectrum and energy level schematic for Nile Blue showing excitation at 615 nm and probe tunable around 660 nm.

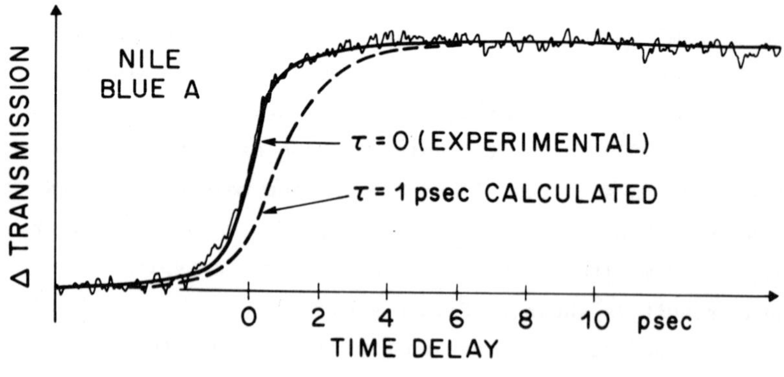

FIGURE 4.6 Apparently instantaneous rise in gain for Nile Blue at 660 nm.

relaxed, Stokes-shifted gain spectrum appears within the resolution of the measurement.

Several explanations are possible for this rapid spectral relaxation. Intramolecular relaxation of the vibrational energy may occur, distributing the energy over a large number of modes or in such a way that the electronic transition is no longer effected. Some intermolecular transfer to solvent molecules may also take place on this timescale. Perhaps most likely is that, for the number and variety of modes involved, there may be very little dependence of the electronic spectrum on the vibronic population distribution. What remains, however, is that a gain spectrum indistinguishable from that of the thermally relaxed molecules develops on a subpicosecond timescale. Similar results were obtained for spectral relaxation within the ground state following subpicosecond pulse stimulated emission to an excited vibrational level of the ground state (Wiesenfeld and Ippen, 1979).

Very recently vibrational relaxation has been resolved in a Li F_a(II) color center in KCℓ at low temperature (Wiesenfeld et al., 1981). Results obtained in these experiments are shown in Figure 4.7. The data at 15°K and lower show a 10 psec delay in the appearance of gain at 2.62 μm (tunable probe) following excitation with a subpicosecond pulse at 615 nm. At higher temperatures, this delay becomes less in a manner consistent with a multiphonon relaxation process.

Another interesting study of nonradiative relaxation has been made of azulene, a molecule of theoretical interest because of its unusual fluorescence properties. The molecular structure and a schematic energy diagram of azulene is given in Figure 4.8. Of particular interest is the lifetime of the lowest excited single state, S_1 (Rentzepis, 1968). A very low quantum efficiency of fluorescence indicates that this level is rapidly depopulated by nonradiative decay. In this case, the fact that flourescence can be observed following excitation to high states suggests a method for monitoring the population in S_1.

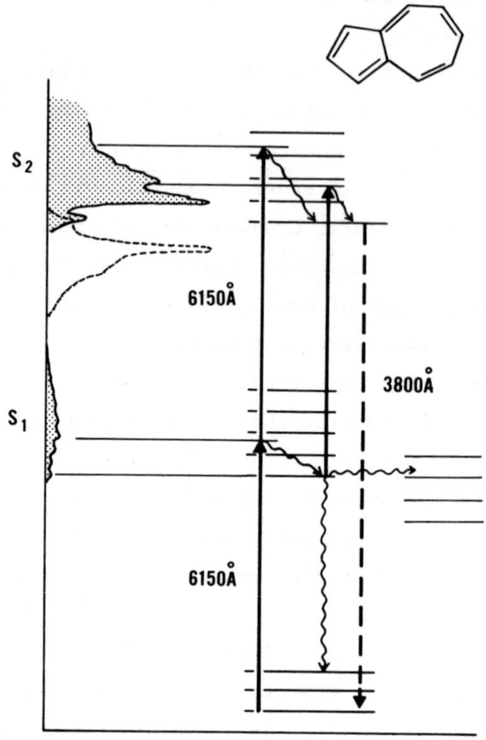

FIGURE 4.8 Structure and energy diagram of the molecule azulene.

The experimental arrangement of Figure 3.3 lends itself well to this purpose. With the KDP in place, ultraviolet light is detected only when the two pulses arrive at the same time. If the KDP is replaced by a thin cell containing a solution of azulene, the situation is different. A single pulse can excite molecules to S_1 where they remain for some short time τ. If the second (delayed) pulse arrives within this time, some molecules are elevated further to a state from which fluorescence can occur. Each individual pulse can also induce two-step excitations but this only adds a constant background independent of delay. By measuring the average fluorescence as a function of delay, one can deduce τ. A reference curve (one for which τ =

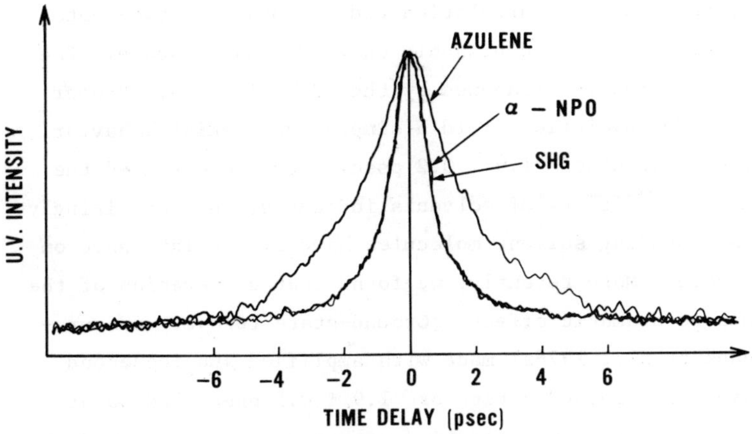

FIGURE 4.9 Measurement of the lifetime of S_1 in azulene. Time-resolved, two-step-induced fluorescence is compared with an instantaneous response.

= 0) is obtained by substituting a dye (α-NPO) that has no intermediate state; the experiment then becomes simply an auto-correlation measurement of the pulses by two-photon fluorescence.

Typical experimental curves we obtained in this study appear in Figure 4.9. Two reference curves (τ = 0), one by a second-

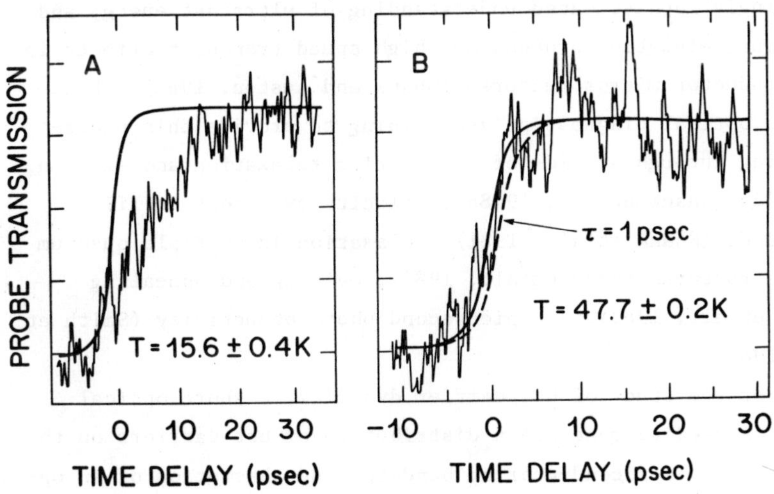

FIGURE 4.7 Picosecond onset of gain in an F_A(II) color center at two temperatures.

harmonic-generation autocorrelation and the other by two-photon fluorescence, are indistinguishable on a 0.1 psec scale. The third curve is clearly broadened by the S_1 lifetime. Proper analysis and deconvolution yield a simple exponential behavior with a time constant of 1.9 ± 0.2 psec. That we obtained the same result in a variety of solvents indicates, not surprisingly, that the surrounding solvent molecules have little influence on this timescale. More recently, we found that deuteration of the azulene has no dramatic effect. Ground-state recovery measurements (Shank et al., 1978a) made with amplified subpicosecond pulses have also yielded a time of 1.9 ± 0.5 psec and point to the conclusion that the lifetime of state S_1 is limited by direct coupling to the ground state. These experiments and others like them provide a basis for comparing theoretical models and will, we hope, lead to a better understanding of nonradiative relaxation.

4.3 Semiconductor Physics

Advances in semiconductor technology are requiring an increasingly sophisticated understanding of ultrafast energy and momentum relaxation process and high speed transport effects in semiconductor microstructures (Shank and Auston, 1982). Picosecond optical diagnostics are helping to provide this understanding through studies of hot carrier relaxation and exciton dynamics (Shank et al., 1978b), velocity overshoot in III-V compounds (Shank et al., 1981), relaxation in multiple quantum well structures (Fork et al., 1980), melting and annealing (Kurz et al., 1983), and picosecond photoconductivity (Smith et al., 1981a).

For the study of hot carrier dynamics, a short optical pulse is used to generate a distribution of hot carriers on the surface of a semiconductor by band-to-band absorption of an optical pulse of frequency greater than the bandgap. A second, less intense pulse then probed the incremental reflectivity change at a delayed time. As the carrier distribution relaxes by optical phonon emission, the reflectivity is altered due to the continuously changing occupation probabilities of the different electronic states until a carrier distribution is in thermal equilibrium with the lattice. Details of the relationship between the optical reflectivity and the carrier distribution depend on the frequency of the probing pulse and the band structure of the particular semiconductor.

In Figure 4.10 we plot the results (Shank et al., 1978c) for the case of gallium arsenide excited with a short pulse at 4 eV and probed with a 2 eV pulse. In this example the sign of the reflectivity change goes from negative to positive as the carriers relax through the probing pulse energy. The energy loss rate for hot carriers can be estimated from the zero crossing to be .4 eV/psec.

A more complete picture of the dynamics of optically induced changes in GaAs near the bandgap has been obtained by subpicosecond continuum probing (Shank et al., 1978b). Amplified picosecond pulses are divided into two beams with variable relative delay. One beam of pulses is converted to 750 nm by stimulated Raman scattering in ethanol. These pulses excite the thin (0.5

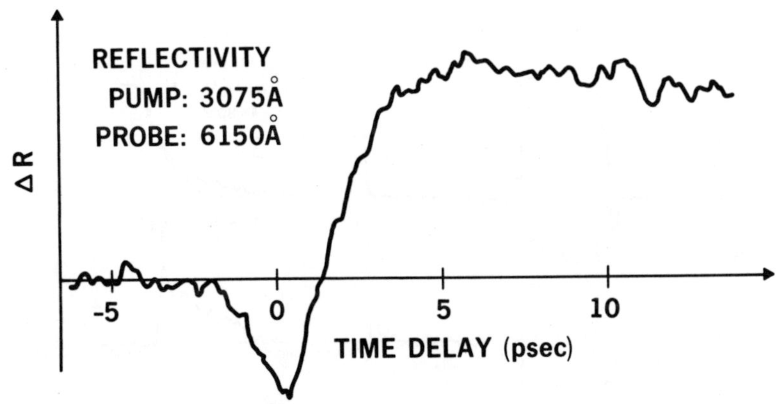

FIGURE 4.10 Reflectivity dynamics of GaAs due to relaxation of hot carriers.

μm), cooled (80°K) sample of GaAs. The second beam is used to generate the continuum probe. Filters attenuate and select the wavelength range of interest (780 nm - 840 nm) to probe near the band edge of GaAs. This spectrum, after transmission through the sample, is dispersed by a spectrometer onto the face of an optical multichannel analyzer (OMA). The complete spectrum may then be registered and averaged over many pulses at each point of delay. Transmission at selected wavelengths was also measured as a continuous function of delay.

The development of the transmission spectrum with time is illustrated selectively in Figure 4.11. Before the 750 nm pump pulse arrives ($\tau < 0$) one sees the unperturbed band edge absorption as well as a sharp absorption peak due to free excitons. Immediately following excitation ($\tau = 1$ psec) the exciton resonance is no longer evident. This may be attributed to screening of the exciton resonance by the excited carriers (Shah et al., 1977). At the same time the band edge has shifted to just below the exciton peak. The spectrum taken at $\tau = 12$ psec shows the effects of band filling after carriers have come into equilibrium with the lattice. Intermediate measurements reveal the time constant of this equilibration to be about 4 psec.

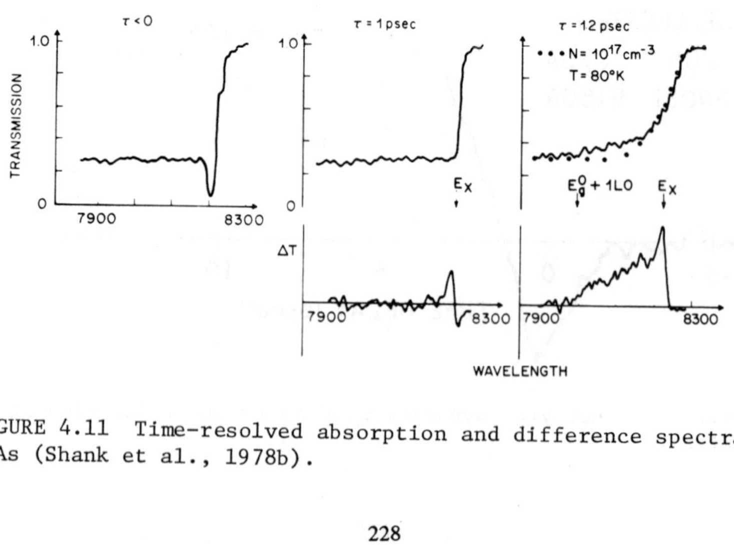

FIGURE 4.11 Time-resolved absorption and difference spectra of GaAs (Shank et al., 1978b).

Saturation and response of the exciton is of practical interest for the development of optical switching elements (Gibbs et al., 1979). Utility of excitonic features at room temperature has recently been made possible by multiple quantum well structures (Miller et al., 1982).

4.4 Optical Electronics

In parallel with these developments have come equally rapid advances in the area of optical electronics and detection. High speed photoconductive switches, samplers, and mixers have been fabricated with a variety of materials and studied with a view toward different applications. The fastest sampling times and the shortest electrical pulses have been achieved with amorphous silicon films and radiation damaged silicon-on-sapphire (SOS) films (Smith et al., 1981a). An advantage of radiation damaged material is that the speed-sensitivity trade-off can be optimized for specific applications. Response times as short as 8 psec have been achieved while maintaining a sensitivity 10 times greater than that of amorphous material.

New, high-speed photoconductive detectors have recently been reported for use at long wavelengths. Detectors of p-type InGaAs (Degani et al., 1981) with a bandgap of 0.75 eV have yielded responses as fast as 80 psec (FWHM) to pulses at 1.3 μm. These devices take advantage of the high drift velocity of electrons as the minority carriers and have a photo-response extending to 1.7 μm. A partially ordered germanium (ρ-Ge) material has also been used for the same purpose (DeFonzo et al., 1981). The material is formed by laser irradiation of amorphous Ge under electrical bias to improve sensitivity while still maintaining a picosecond lifetime. A detector response of less than 50 psec has been observed and photoresponse to 1.7 μm is expected.

A novel and exciting application of the short electrical pulses is the direct probing of high-speed solid-state electronic

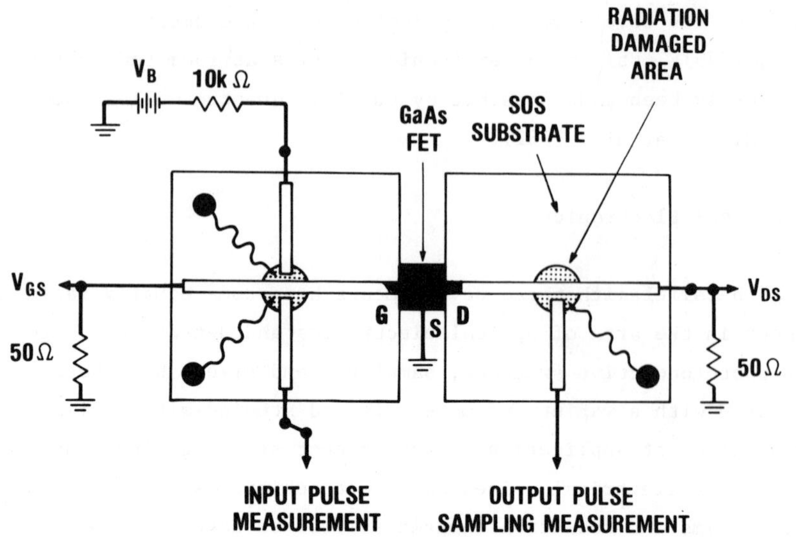

FIGURE 4.12 Schematic of a picosecond electronic circuit used to measure the response of a GaAs FET (Smith et al., 1981b)

devices. In the first experiments of this kind (Smith et al., 1981b), electronic response times of a GaAs microwave FET have been studied with a temporal resolution of less than 10 psec. Figure 4.12 illustrates the technique. Picosecond optical pulses, indicated by the wavy lines, are focused on gaps in microstrip transmission lines where very high speed photoconductors were made by damaging silicon-on-sapphire wafers with radiation. Two pulses on the left half of the figure activate the gate of the FET. The current from the drain is then sampled by the third optical pulse on the right. Their particular results showed both a fast and a slow component to the FET response. Further experimentation should provide new insight into the high-speed operation of such devices. The conversion of optical to electrical pulses will also facilitate the probing of sub-micron structures that are smaller than the minimum focus of an optical beam. So far, most experiments in this area have utilized modelocked dye lasers. With some improvement in gap sensitivity it may soon be possible to use compact, picosecond diode lasers for this type of diagnostics.

REFERENCES

Auston, D.H., 1975, Appl. Phys. Lett. 26, 101.

Auston, D.H., 1977, *Ultrashort Light Pulses*, edited by S.L. Shapiro (Springer Verlag), chapter 4.

DeFonzo, A.P., 1981, Appl. Phys. Lett. 39, 480.

Degani, J., R.F. Leheny, R.E. Nahory, M.A. Pollack, J.P. Heritage, and J.C. DeWinter, 1981, Appl. Phys. Lett. 38, 27.

Fork, R.L., C.V. Shank, and B.I. Greene, 1980, *Picosecond Phenomena II* (Springer Verlag), p. 280.

Gibbs, H.M., S.L. McCall, T.N.C. Venkatesan, A.C. Gossard, A. Passner, and W. Wiegman, 1979, Appl. Phys. Lett. 35, 451.

Hesselink, W.H., and D.A. Wiersma, 1978, Chem. Phys. Lett. 56, 227.

Ippen, E.P., C.V. Shank, and A. Bergman, 1976, Chem. Phys. Lett. 38, 611.

Kurz, H., J.M. Liu, R. Yen and N. Bloembergen, 1983, J. Quant. Electron., to be published.

Laubereau, A., D. von der Linde, and W. Kaiser, 1972, Phys. Rev. Lett. 28, 1162.

Miller, D.A.B., D.S. Chemla, and P.W. Smith, 1982, Appl. Phys. B28, 96 (Proc. 12th IQEC)

Olson, R.W., F.G. Patterson, H.W.H. Lee, and M.D. Fayer, 1981, Chem. Phys. Lett. 77, 403.

Rentzepis, P.M., 1968, Chem. Phys. Lett. 2, 117.

Shah, J., R.F. Leheny, and W. Weigman, 1977, Phys. Rev. B15, 1577.

Shank, C.V. and D.H. Auston, 1982, Science 215, 797.

Shank, C.V. and E.P. Ippen, 1975, Appl. Phys. Lett. 26, 62.

Shank, C.V., E.P. Ippen, O. Teschke, and R.L. Fork, 1978a, Chem. Phys. Lett. 57, 433.

Shank, C.V., R.L. Fork, R.F. Leheny, and J. Shah, 1978b, Phys. Rev. Lett. 42, 112.

Shank, C.V., D.H. Auston, E.P. Ippen, and O. Teschke, 1978c, Solid State Commun. 26, 567.

Shank, C.V., R.L. Fork, B.I. Greene, F.K. Reinhardt, and R.A. Logan, 1981, Appl. Phys. Lett. 38, 104.

Smith, P.R., D.H. Auston, A.M. Johnson, and W.M. Augustiniak, 1981a, Appl. Phys. Lett. 38, 47.

Smith, P.R., D.H. Auston and A.M. Augustyniak, 1981b, Appl. Phys. Lett. 39, 739.

Sundstrom, V., T. Gillbro, and H. Bergstrom, 1982, Chem. Phys.

Waldeck, D., A.J. Cross, Jr., D.B. McDonald, and G.R. Fleming, 1981, J. Chem. Phys. 74, 3381.

Wiesenfeld, J.M., and E.P. Ippen, 1979, Chem. Phys. Lett. 67, 213.

Wiesenfeld, J.M., L.F. Mollenauer, and E.P. Ippen, 1981, Phys. Rev. Lett. 47, 1968.

Yu, W., F. Pelligrino, M. Grant, and R.R. Alfano, 1977, J. Chem. Phys. 67, 1766.

5. PICOSECOND SEMICONDUCTOR LASERS

5.1 Introduction

Interest in the generation of ultrashort optical pulses with semiconductor lasers has been increasing rapidly within the past several years. For scientific applications, this interest has been motivated by the potential for tunable picosecond lasers at wavelengths not easily attainable with other sources such as dye lasers. For high speed optical communications and ultrafast signal processing the possibility of compact reliable sources of ultrashort pulses is very promising, especially at wavelengths compatible with distortion-free transmission over optical fibers. In any case, efforts along these lines can also be expected to test the limits of our understanding of ultrafast and non-linear processes in semiconductors and perhaps even clarify phenomena of importance to other devices.

5.2 Optically pumped semiconductors

Optically pumped semiconductor lasers were among the first lasers to be demonstrated (Phelan and Rediker, 1965; Stillman et al., 1966). Virtually any direct-bandgap semiconductor can be used for this purpose, so that the potential (and already demonstrated) wavelength coverage is considerably greater than that for injection lasers alone. The most widely studied systems have utilized II-VI compounds (Johnson and Holonyak, 1968, Shaklee and Leheny, 1971) with picosecond pulses having been generated only recently (Roxlo and Salour, 1981). With advances in crystal growing techniques, it has also become possible to work successfully with III-V compounds (Burnham et al., 1971; Rossi et al., 1972), and picosecond pulses have been obtained from these materials as well (Duguay et al., 1980).

One currently successful means of optical excitation is synchronous pumping. In a manner very much like that described in Chapter 2, an actively modelocked ion laser pumps a thin semiconductor crystal in a resonator of matched length. Pulses 4-10 psec in duration have been obtained in the visible at a repetition of 80 MHz from platelets of CdS and CdSSe (Roxlo and Salour, 1981; Salour, 1982). With a modelocked Kr^+ laser pumping thin LPE layers of InGaAsP, tunable bandwidth-limited pulses of about 6 psec duration in the range 1.1 to 1.3 μm have also been achieved (Putnam et al., 1982; Salour, 1982). This type of laser has the desirable features of being able to cover broad spectral ranges with a single pumping source and of being mechanically stable (i.e. no flowing dye). Material composition could be varied as a function of position or platelets interchanged rapidly. The primary disadvantages to be overcome are average power limitation due to problems of heat dissipation and the necessity of operating at low temperature.

Another, very direct, optical pumping scheme utilizes thin semiconductor samples in an ultrashort (less than 1 μm) resonator (Duguay et al., 1980; Damen et al., 1981). Excitation is with a single, ultrashort pulse from a dye laser. Oscillation at the semiconductor laser wavelength builds up rapidly in the ultrashort resonator and then depletes the gain to terminate lasing within picoseconds. Broad bandwidths can be achieved because the intense pumping pulse creates electron-hole pair densities on the order of $10^{20}/cm^3$ and raises the Fermi level high into the conduction band. Tuning over a wide range is made possible by translating the slightly wedged resonator. With sub-picosecond pumping of GaAs, pulse durations varied from 3 psec at 7900Å to 8 psec at 8900Å (Damen et al., 1981). Sub-picosecond pulses had been observed with slightly thicker samples earlier (Duguay et al., 1980), and the same scheme has been extended to InGaAsP (Stone et al., 1981).

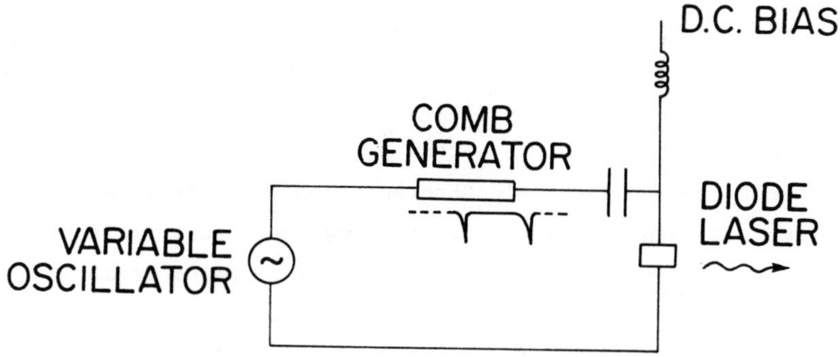

FIGURE 5.1 Simple experimental arrangement for pulse modulating a diode laser

5.3 Modulated Diode Lasers

Direct, current modulation offers perhaps the simplest means for generating short pulses from semiconductor diode lasers. Laser current is modulated either with a sinusoidal RF drive on top of a DC bias (Ito et al., 1979a, 1979b; Aspin et al., 1981) or with short current pulses (Lin et al., 1980; Liu et al., 1981; Torphammer and Eng, 1980). An experimental schematic of such an arrangement is shown in Figure 5.1. In all cases it is possible to generate light pulses considerably shorter than the current pulse duration. The pulse shortening dynamics that result from cavity build-up time and rapid gain depletion have been referred to as gain switching and controlled relaxation oscillation. Relatively straightforward, coupled rate equation analyses have been advanced to explain these dynamics (Ikegami et al., 1970; Adams, 1973; Danielson, 1976) and there seems to be good correspondence between theory and experiment. Pulses in the range 20-60 ps have been generated in this way with both GaAℓAs and InGaAsP diodes at repetition rates from 1MHz to 3GHz (Inaba, 1982). At the low repetition rates, short pulse pumping is clearly advantageous because sinusoidal RF modulation becomes too slow; but for high rates the modulation functions become similar and there is little difference in the pulse generated by the two schemes.

Incorporating saturable absorber elements or active loss modulators within the diode structure may be used to further enhance gain switching. One problem still to be overcome is the strong spectral broadening that occurs when diodes are rapidly modulated. That is, without additional resonator control, the pulses are far from transform-limited. Pulse durations are also limited to be longer than several transit times of the diode cavity. Thus, even with short diodes 10-20 psec would seem to be the limit. Figure 5.2 illustrates that pulse generation by a uniformly pumped, isolated diode produces pulses that are longer than the length of the diode; but that in an external resonator, the diode can be a distributed medium that forms pulses shorter than its own length.

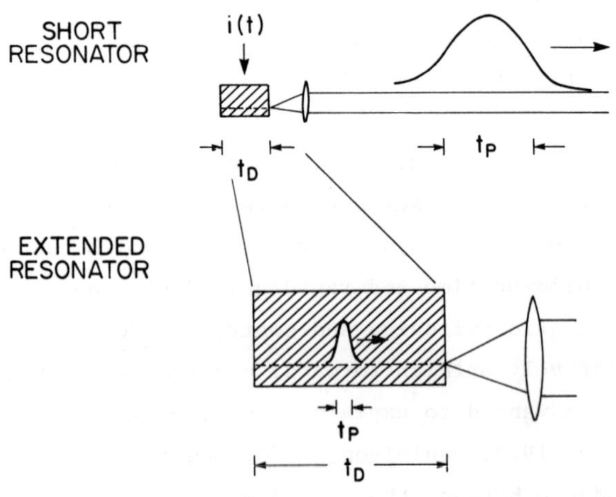

FIGURE 5.2 Pulse generation by a lumped element vs. pulse generation in an extended resonator.

5.4 Modelocked Diode Lasers

If a lesson is to be learned from modelocked dye lasers (Chapter 2), it is that the shortest pulses can be obtained when the dynamic response time of the gain medium is on the order of a round-trip time of the laser resonator. In the case of a semiconductor diode laser, where relaxation is in the nanosecond range, this requires that the diode be operated in an extended resonator. The first experiments to demonstrate the efficacy of this theory used a cw GaAℓAs diode laser at the focus of an external convex mirror which formed a cavity about 5 cm long (Ho et al., 1978). In addition, the diode current was modulated at a frequency of approximately 3GHz, corresponding to the inverse of the cavity round-trip time. Pulses 23 psec in duration were obtained by SHG in $LiIO_3$ assuming a Gaussian pulse shape. Soon thereafter, similar active modelocking results were obtained with a cw InGaAsP diode operating at 1.21 μm (Glasser, 1978) and in GaAℓAs by other workers (van der Ziel and Mikulyak, 1980).

Spectral broading during modelocking in diode lasers tends to produce independent oscillation on clusters of external cavity modes centered about different longitudinal modes of the short diode. This results in pulses that are not bandwidth limited and that produce autocorrelation traces such as those shown at the top of Figure 5.3. The narrow coherence spikes reflect the broad spectral width, and their periodicity is a result of multiple reflections within the diode (or, in the frequency domain, of the diode longitudinal mode structure). This sub-structure can be removed by restricting the bandwidth of oscillation with an etalon (Ho, 1979; Ito et al., 1979a; Ippen et al., 1980a) or by cleaving the end facet of the diode at an angle (Holbrook et al., 1980). What is perhaps unexpected is that modelocking can even improve somewhat and the pulses become shorter as the bandwidth is restricted. This was noted for

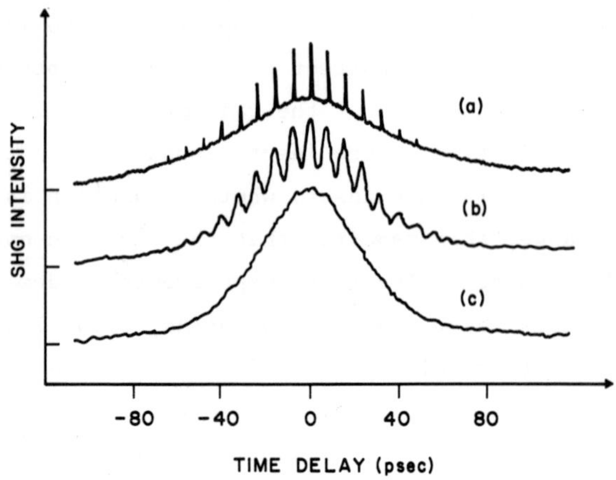

FIGURE 5.3 Autocorrelation measurements of pulses with (a) no bandwidth control (b) partial bandwidth control and (c) an optimized etalon.

a passively modelocked system and is illustrated in Figure 5.3 (Ippen et al., 1980a). Clearly, if the bandwidth is restricted too much the pulse duration will again broaden. Excess bandwidth in the laser may result in pulse broadening due to spontaneous emission noise (Haus and Ho, 1979) or to group velocity differences between clusters.

The shortest diode laser pulses have been achieved by passive modelocking. It was observed early that diodes in extended resonators were susceptible to self-pulsing (Glasser, 1978). Shortly thereafter, a specific study of passive modelocking produced the first pulses shorter than a round-trip of the diode element itself (Ippen et al., 1980a). The experimental set-up is illustrated in Figure 5.4. The resonator consists of a 40x microscope objective, a variable-gap air-spaced etalon (which was removed to obtain the shortest pulses) and a flat mirror. Output pulse characteristics were

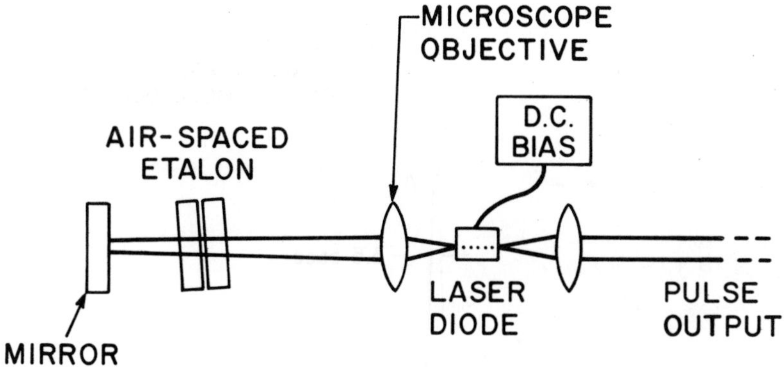

FIGURE 5.4 External resonator configuration for passive modelocking of diode lasers.

found to depend very little upon which of the two critical lens positions were used (except that the one that images a waist on the flat mirror is mechanically more stable) and upon cavity length over a range of 7–20 cm. Dramatic pulse shortening was observed after aging some of the diodes until an increase in threshold was observed. The shortest single pulse (i.e. one per round-trip) reported so far in the literature is 5.1 psec produced in this manner (Ippen et al., 1980a). A slightly shorter one, also obtained with an aged diode, is shown in Figure 5.5.

One can infer from passive modelocking theory (Chapter 1) and the above results that the observed increase in threshold is associated with the introduction of an additional absorption loss, that this additional loss is saturable, and that it saturates more easily than the laser gain. Several proposed mechanisms (Dixon and Joyce, 1979; Copeland, 1978) could satisfy these conditions.

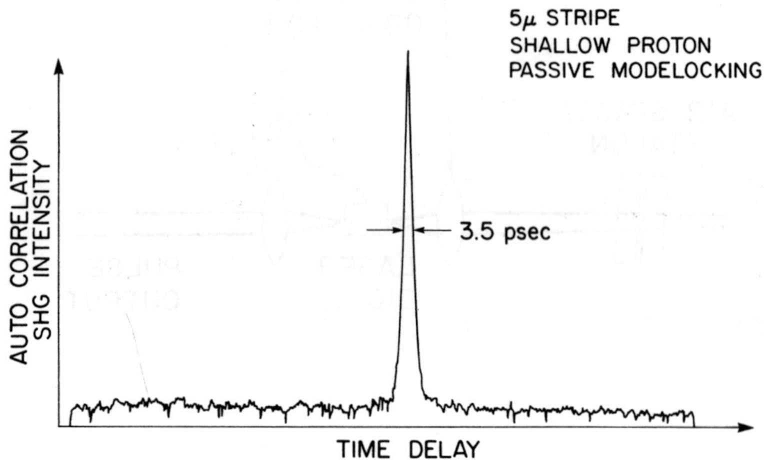

FIGURE 5.5 SHG autocorrelation measurement of repetitive, ultrashort, single pulse output of diode. Pulsewidth indicated assumes a sech^2 shape.

In order to better understand the aging process, and to understand why some diodes produced shorter pulses than others, we monitored side emission from the active stripe by viewing through the top of the diodes with an infrared microscope (Ippen et al., 1980b). With current confinement by proton bombardment the top electrode metallization can be separated at the center to allow such viewing (Hartman et al., 1979). Concomitant with increases in threshold we observed the appearance and spreading of dark areas in the stripe emission. Interestingly, the shortest pulses (less than 6 psec) were always produced when the dark area formed near the face of the diode that served as the end of the resonator. Such an experimental observation is reproduced schematically in Figure 5.6. A reason for such correlation may be improved saturability of an absorber under these conditions.

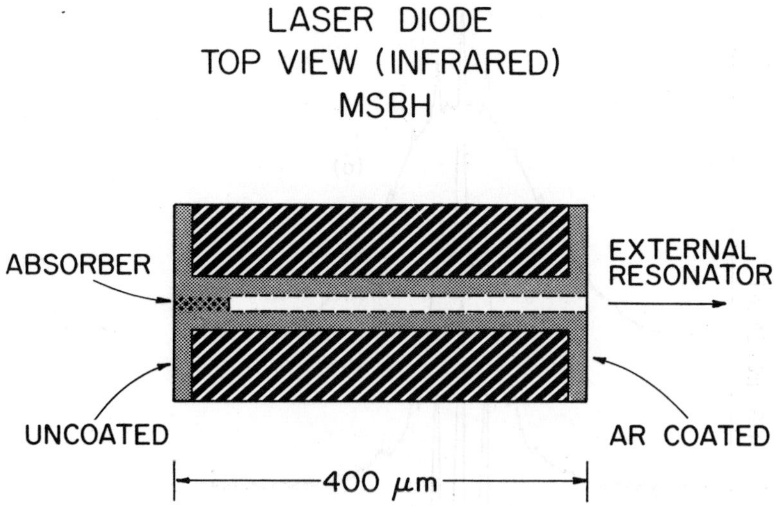

FIGURE 5.6 Schematic representation of view through top of MSBH diode laser with separated electrodes. The (dark area) absorbing region is on the left at one end of the extended resonator

An optical pulse shorter than the diode only produces a standing-wave field near the end face. Such a field results in the coherent coupling between forward and backward wave as discussed in Chapters 1 and 2.

Occasionally, another mode of behavior is observed during the aging process. The time development is shown in Figure 5.7 with the end result in Figure 5.8. Here output consists of a short burst of pulses, each 1.3 psec in duration (Ippen et al., 1980b), with separation about equal to a diode round-trip time. In our experience this has often, but not always, been due to growth of the absorber at the other (interior to resonator) end of the diode.

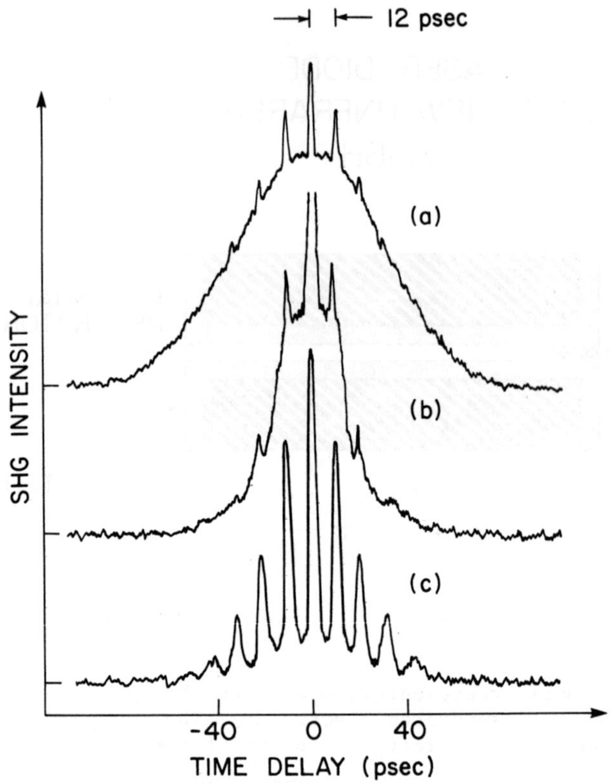

FIGURE 5.7 Time development of autocorrelation traces during aging that produces multiple ultrashort pulses.

Recently the observation of sub-picosecond pulses has been reported for similar behavior in aged diodes (Yokoyama et al., 1982) and with diodes into which a saturable absorption was specifically introduced at the exit facet of the diode by proton bombardment (van der Ziel et al., 1981). An absorbing region of 8-10 μm in depth, produced with 600 keV protons, provided the necessary nonlinearity in the buried optical guide laser. This latter work is certainly an important step toward controlling the material nonlinearities necessary for ultrashort pulse generation. Further work will hopefully show that this technique, or some other, can be utilized reproducibly, and in a

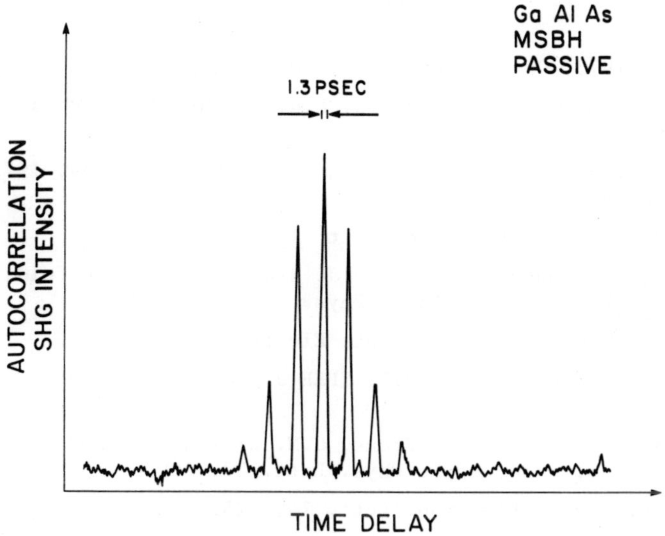

FIGURE 5.8 Autocorrelation trace of multiple picosecond pulse output from GaAℓAs diode laser.

way that is consistent with long device life.

Now that picosecond and sub-picosecond pulse generation with diode lasers has been demonstrated, these results will certainly be pursued into the wavelength regime where such pulses can propagate with minimum dispersion in optical fibers. Such a potential capability also then calls out for the invention of novel techniques for the utilization of such pulses.

The author wishes to acknowledge collaboration with D.J. Eilenberger on the previously unpublished diode laser work described above.

REFERENCES

Adams, M.J., 1973, Opto-electron. 5, 201.

Aspin, G.J., J.E. Carroll, and R.G. Plumb, 1981, Appl. Phys. Lett. 39, 860.

Burnham, R.D., N. Holonyak, D.L. Keune, and D.R. Scifres, 1971, Appl. Phys. Lett. 18, 160.

Copeland, J.A., 1978, Electron. Lett. 14, 309.

Damen, T.C., M.A. Duguay, J. Shah, J. Stone, J.M. Wiesenfeld, and R.A. Logan, 1981, Appl. Phys. Lett. 39, 142.

Danielson, M., 1976, IEEE J. Quant. Electron. QE-12, 657.

Dixon, R.W. and W.B. Joyce, 1979, IEEE J. Quant. Electron. QE-15, 470.

Duguay, M.A., T.C. Damen, J. Stone, J.M. Wiesenfeld, and C.A. Burrus, 1980, Appl. Phys. Lett. 37, 369.

Glasser, L.A., 1978, Electron. Lett. 14, 725.

Hartman, R.L., R.A. Logan, L.A. Koszi, and W.T. Tsang, 1979, J. Appl. Phys. 50, 4616.

Haus, H.A., and P.T. Ho, 1979, IEEE J. Quant. Electron. QE-15, 1258.

Ho, P.T., L.A. Glasser, E.P. Ippen, and H.A. Haus, 1978, Appl. Phys. 33, 241.

Ho, P.T., 1979, Electron. Lett. 15, 526.

Holbrook, M.B., W.E. Sleat, and D.J. Bradley, 1980, Appl. Phys. Lett. 37, 59.

Ikegami, T., K. Kobayashi and Y. Suematsu, 1970, Electron. and Commun. Japan 53-B, 82.

Inaba, H., 1982, Jap. Ann. Rev. Electron. Comput. and Telecomm., Opt. Devices and Fibers, 60.

Ippen, E.P., D.J. Eilenberger, and R.W. Dixon, 1980a, Appl. Phys. Lett. 37, 267.

Ippen, E.P., E.J. Eilenberger, R.W. Dixon, 1980b, *Picosecond Phenomena II*, edited by Hochstraser et. al. (Springer Verlag), 21.

Ito, H., H. Yokoyama, S. Murata, and H. Inaba, 1979a, Electron. Lett. 15, 738.

Ito, H., H. Yokoyama, S. Murata, and H. Inaba, 1979b, IEEE J. Quant. Electron. QE-17, 663.

Johnson, M.R., and N. Holonyak, 1968, J. Appl. Phys. 19, 3977.

Lin, C., P.L. Liu, T.C. Damen, D.J. Eilenberger, and R.L. Hartman, 1980, Electron. Lett. 16, 600.

REFERENCES (continued)

Liu, P.L., C. Lin, I.P. Kaminow, and J.J. Hsieh, 1981, IEEE J. Quant. Electron. QE-17, 671.

Phelan, R.J., and R.H. Rediker, 1965, Appl. Phys. Lett. 6, 70.

Putnam, R.S., C.B. Roxlo, M.M. Salour, S.H. Groves, and M. Plonko, 1982, to be published.

Rossi, J.A., S.R. Chinn, and A. Mooradian, 1972, Appl. Phys. Lett. 20, 84.

Roxlo, C.B. and M.M. Salour, 1981, Appl. Phys. Lett. 38, 738.

Salour, M.M., 1982, *Picosecond Phenomena III* (Springer Verlag).

Shaklee, K.L. and R.F. Leheny, 1971, Appl. Phys. Lett. 18, 475.

Stillman, G.E., M.D. Sirkis, J.A. Rossi, M.R. Johnson, and N. Holonyak, 1966, Appl. Phys. Lett. 9, 268.

Stone, J., J.M. Wiesenfeld, A.G. Dentai, T.C. Damen, M.A. Duguay, T.Y. Chang, and E.A. Caridi, 1981, Opt. Lett. 6, 534.

Torphammer, P.T., and S.T. Eng, 1980, Electron. Lett. 16, 587.

Van der Ziel, J.P., and R.M. Mikulyak, 1980, J. Appl. Phys. 51, 3033.

Van der Ziel, J.P., W.T. Tsang, R.A. Logan, R.M. Mikulyak, and W.M. Augustyniak, 1981, Appl. Phys. Lett. 39, 525.

Yokoyama, H., H. Ito, and H. Inaba, 1982, Appl. Phys. Lett. 40, 105.

REFERENCES (continued)

SEMICONDUCTOR LIGHT SOURCES AND DETECTORS

Michiharu Nakamura
Central Research Laboratory, Hitachi Ltd.
Kokubunji, Tokyo 185, Japan

ABSTRACT

A review is presented on semiconductor lasers and detectors which have been developed extensively for communications and other applications. The wavelength regions mainly discussed here are 0.7-0.9 μm and 1.2-1.6 μm. Detailed characteristics of the devices and their fabrication technology are summarized.

TABLE OF CONTENTS

1. SEMICONDUCTOR LIGHT SOURCES: DEVICE PHYSICS

1.1 Introduction

The magnificent technical progress of the last two decades owes much to the ceaseless progress of Si IC technology. Si devices are now penetrating into all fields of human life, not only in computers, but also in communication equipment, consumer products, etc. However a new technical tide has also been growing in the same decades. It is <u>optoelectronics</u>. It is still premature in all aspects, but is expected to be a key technology in the future.

Among the most important devices supporting the recent optoelectronics are semiconductor light sources and detectors. The utilization of semiconductor materials makes it possible to realize high efficiency, small-sized, cheap components. These are essential features for optoelectronics to become practical.

In table 1.1 is summarized the state of art of light sources. Light emitting diodes (LEDs) and laser diodes of GaAlAs/GaAs and InGaAsP/InP are the well developed light sources which cover 0.75-0.89 μm and 1.2-1.6 μm, respectively. High speed detection of light is achieved by PIN photodiodes and avalanche photodiodes (APDs) of Si (short wavelength region), Ge and InGaAs(P) (long wavelength region). In this chapter, the device physics of semiconductor light sources are described.

1.2 Birth of Semiconductor Light Sources

The first observations of the stimulated emission of light from GaAs p-n junction diodes were reported in 1962 from GE (Hall et al., 1962), IBM (Nathan et al., 1962), and MIT Lincoln Laboratory (Quist et al., 1962). The diodes were operated at 4.2 K and 77 K, showing threshold behavior in light intensity-current curves and also spectral narrowing. Since then much attention

Table 1.1 Typical characteristics of semiconductor light sources.

Wavelength	Material	Power	f_c
0.75-0.89 μm	GaAlAs/GaAs	10-100 μW* (LED)	0.1 GHz (LED)
1.2-1.6 μm	InGaAsP/InP	5-20 mW (Laser)	1 GHz (Laser)

* power coupled into a fiber

f_c: cut-off frequency

was focused to this field, but it was in 1970 that cw operation of semiconductor lasers became possible by introduction of a GaAlAs/GaAs double heterostructure. The double heterostructure first suggested by Kroemer (Kroemer, 1963) enabled efficient carrier confinement and optical waveguiding simultaneously, thus leading to low threshold operation of laser diodes (Panish et al., 1970).

The problem encountered just after the first cw operation of GaAlAs/GaAs double heterostructure lasers was the poor reliability of the devices. The origin of the poor reliability was found, after much elaborate work, to be the rapid growth of dislocation networks in the light emitting region. Along with the improvements in laser reliability, transverse mode control has also been a matter of hot arguments. Various waveguide structures, which may be classified into two schemes, gain guiding and index guiding, have been introduced to stabilize the transverse mode in semiconductor lasers (Nakamura et al., 1981).

The double heterostructure has also been used to realize high efficiency LEDs.

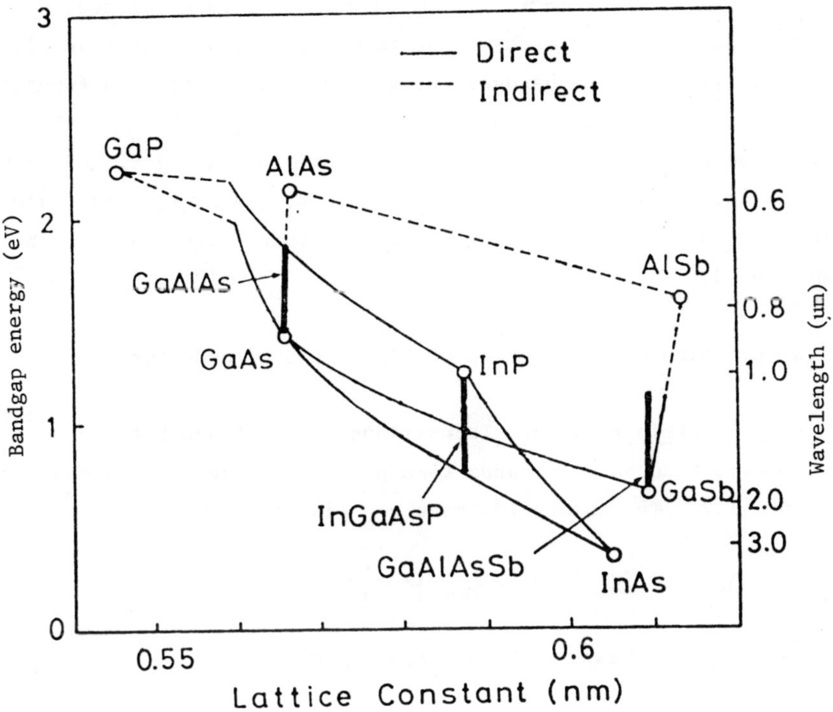

Fig. 1.1 Bandgap energy of III-V compound semiconductors as a function of lattice constant.

1.3 Material Selection

The materials of interest here are III-V compound semiconductors. Most of them have zinc-blend crystalline structure. The basic properties of them are given elsewhere (Casey and Panish, 1978). In order to fabricate LEDs or laser diodes, double heterostructure should be formed on a substrate crystal. Figure 1.1 shows the bandgap energies of III-V alloys as functions of lattice constant.

The substrate crystals employed now are GaAs, InP, and GaSb. Using these crystals, $Ga_{1-x}Al_xAs/GaAs$, $In_{1-x}Ga_xAs_yP_{1-y}/InP$, and $Ga_{1-x}Al_xAs_{1-y}Sb_y/GaSb$ double heterostructure devices have been developed.

1.4 Emission and Absorption of Light in Semiconductors

<u>1.4.1 Einstein relation</u> Transitions of electrons between the conduction and valence bands take place in three ways as shown in Figure 1.2. The absorption rate r_{abs} is given by

$$\begin{aligned} r_{abs} &= B_{12} f_1 (1 - f_2) P(E_{21}) \\ f_1 &= \{\exp[(E_1 - F_1)/kT] + 1\}^{-1} \\ f_2 &= \{\exp[(E_2 - F_2)/kT] + 1\}^{-1} \end{aligned} \tag{1.1}$$

where F_1 and F_2 are the quasi-Fermi levels of electrons, $P(E_{21})$ is the photon density at photon energy E_{21}. In addition to being absorbed, the photons also stimulate the transition of electrons, giving similar photons. The probability of the stimulated emission r_{stim} is;

$$r_{stim} = B_{21} f_2 (1 - f_1) P(E_{21}) \tag{1.2}$$

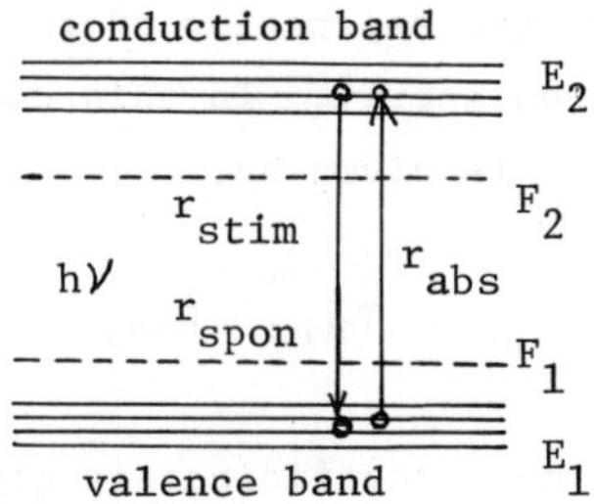

Fig. 1.2 Transitions of electrons between valence band and conduction band.

Also electrons in the conduction band spontaneously return to the valence band at a rate r_{spon};

$$r_{spon} = A_{21} f_2 (1 - f_1) \tag{1.3}$$

At thermal equilibrium,

$$r_{abs} = r_{stim} + r_{spon} , \tag{1.4}$$

$$F_1 = F_2 \tag{1.5}$$

From these equations the transition probabilities B_{12}, B_{21}, and A_{21} are related to each other by the equations

$$B_{12} = B_{21} ,$$
$$A_{21} = (8\pi \bar{n}^3 E_{21}^2 / h^3 c^3) B_{21} . \tag{1.6}$$

where

$\bar{n}$: refractive index

h : Planck's constant

c : light velocity

<u>1.4.2 Necessary condition for stimulated emission of light in semiconductors</u> The necessary condition for the stimulated emission being in excess of absorption is (Bernard and Duraffourg, 1961);

$$B_{21} f_2 (1 - f_1) P(E_{21}) > B_{12} f_1 (1 - f_2) P(E_{21})$$

or $$F_2 - F_1 > E_{21} \tag{1.7}$$

i.e. the separation of the quasi-Fermi levels should exceed the photon energy. In this case, the net stimulated emission rate $\tilde{r}_{stim}$ is given by;

$$\tilde{r}_{stim} = r_{stim} - r_{abs} = B_{12} P(E_{21}) (f_2 - f_1) \tag{1.8}$$

The gain coefficient is given by;

$$g(E_{21}) = \frac{\tilde{r}_{stim}}{P(E_{21}) v_g} = \frac{B_{12}(f_2 - f_1)}{c/\bar{n}} \tag{1.9}$$

where v_g is the group velocity of light.

1.4.3 Transition probability The transition probability B_{21} is given by;

$$B_{12} = (\pi q^2 \hbar / m_e^2 \varepsilon_0 \bar{n}^2 h\nu) \, |\langle \psi_1^*(r) | P | \psi_2(r) \rangle|^2 \tag{1.10}$$

where $\psi_1(r), \psi_2(r)$ are the electron wave functions in the conduction and valence bands, p is the momentum operator, q is the electronic charge, m_e is the electron mass, ε is the dielectric constant, ν is the optical frequency, $\hbar = h/2\pi$.

1.5 Laser Threshold

Semiconductor lasers have two parallel mirrors to form a Fabry-Perot etalon. The oscillation conditions are

$$r_1 r_2 \exp(-2\Gamma L) = 1$$

$$\Gamma = j(\bar{n} + j\bar{k}) K_o \tag{1.11}$$

$$K_o = 2\pi/\lambda, \quad \bar{k} = (\alpha - g)\lambda/4\pi$$

where λ is laser wavelength, L is the laser length, r_1, r_2 are the mirror reflectivities and α is the loss. From the above equations one gets

$$g_{th} = \alpha + \frac{1}{L} \ln (1/r_1 r_2) \tag{1.12}$$

and

$$\lambda = 2L\bar{n}/m_\ell \qquad (m_\ell = 1, 2, \cdots) \tag{1.13}$$

The longitudinal mode spacing $\Delta\lambda$ is

$$\Delta\lambda = \frac{\lambda^2}{2\bar{n}L(1-(\lambda/\bar{n})(d\bar{n}/d\lambda))} \tag{1.14}$$

The current density below threshold is related to r_{spon} by

$$J = q\int_0^\infty r_{spon}(E)\, dE \tag{1.15}$$

Experimentally the following expression is more often used

$$J = J_0 + (g_{max}/\beta)^{1/m} \tag{1.16}$$

where g_{max} is the maximum gain and m is 1-2.

1.6 Waveguide by Double Heterostructure

1.6.1 Waveguide modes Suppose a wave-guiding layer II is embedded between clad layers I and III (Fig. 1.3). The refractive indices of the layers are $\bar{n}_1$, $\bar{n}_2$ and $\bar{n}_3$ respectively. Optical wave-guiding occurs by total reflection of light if $\bar{n}_2 > \bar{n}_1, \bar{n}_3$.

The electromagnetic wave propagating in such a waveguide is expressed by

$$E_y(x) = \begin{cases} C\exp(-\gamma x) & (x \geq 0) \\ C[\cos(\kappa x) - (\gamma/\kappa)\sin(\kappa x)] & (0 \geq x \geq -d) \\ C[\cos(\kappa d) + (\gamma/\kappa)\sin(\kappa d)] \times \exp[\gamma(x+d)] & (-d > x) \end{cases} \tag{1.17}$$

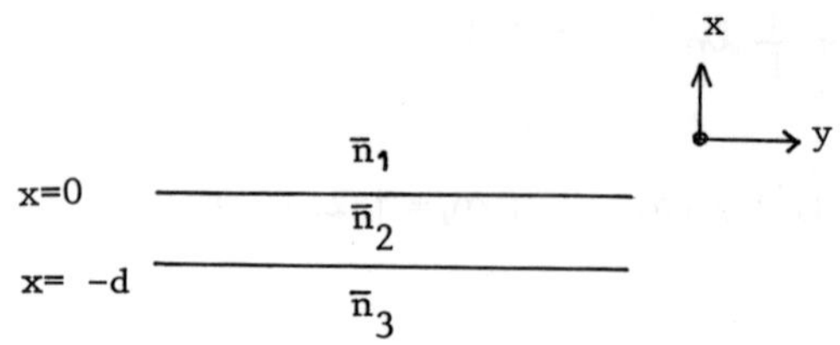

Fig. 1.3 Concept of a dielectric slab waveguide.

for the TE mode. In the above equations, $\bar{n}_2=\bar{n}_3$ was assumed. The parameters, κ etc. are obtained from

$$\tan(\kappa d) = 2\kappa\gamma/(\kappa^2 - \gamma^2)$$

$$\gamma = (\beta^2 - \bar{n}_1^2 k_0^2)^{1/2} \tag{1.18}$$

$$\kappa = (\bar{n}_2^2 k_0^2 - \beta^2)^{1/2}$$

The cut-off condition for such a symmetrical waveguide is given by

$$d < d_c = \tilde{m}\lambda/2(\bar{n}_2^2 - \bar{n}_1^2)^{1/2} \tag{1.19}$$

for the m-th mode. In semiconductor lasers, d=0.05–0.2μm, thus only a fundamental mode (m=0) can propagate.

<u>1.6.2 Refractive index of semiconductors</u> The refractive index of a III-V alloy is a function of the alloy composition x as well as the photon energy. Figure 1.4 shows the refractive index of GaAlAs as functions of photon energy. The dependence of $\bar{n}$ on x is roughly expressed by

$$\bar{n}(x) = 3.590 - 0.710x + 0.091x^2 \tag{1.20}$$

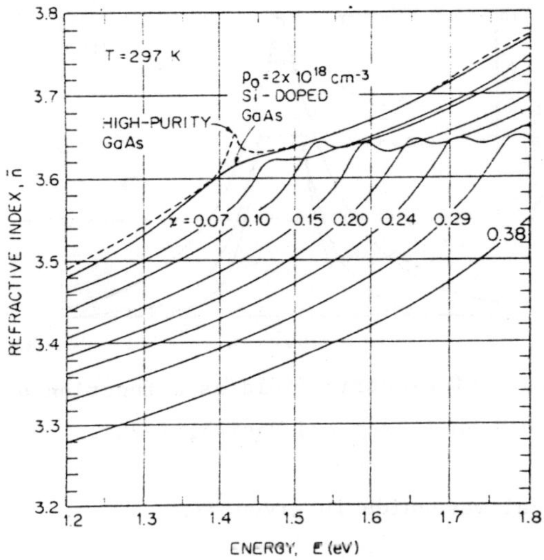

Fig. 1.4 Refractive index as a function of photon energy for $Ga_{1-x}Al_xAs$ (Casey et al. 1974).

at $h\nu$ = 1.38 eV. On $In_{1-x}Ga_xAs_yP_{1-y}$, the refractive index $\bar{n}$ is

$$\bar{n}(y) = 3.40 + 0.256y - 0.059y \tag{1.21}$$

where the quaternary material is assumed to be lattice-matched to InP.

<u>1.6.3 Numerical solution of waveguide modes</u> The square of the electric field gives the optical intensity profile in the waveguide. Calculated intensity profiles are shown for a GaAs-GaAlAs symmetrical waveguide (Casey and Panish, 1978) in Figure 1.5. As is noted the profiles have a considerable amount of optical power outside the waveguide layer.

Since only the electromagnetic fields in the laser active layer interact with the injected carriers, it is useful to define a confinement factor as a ratio of the light intensity within the active layer to the total light intensity. The confinement factor

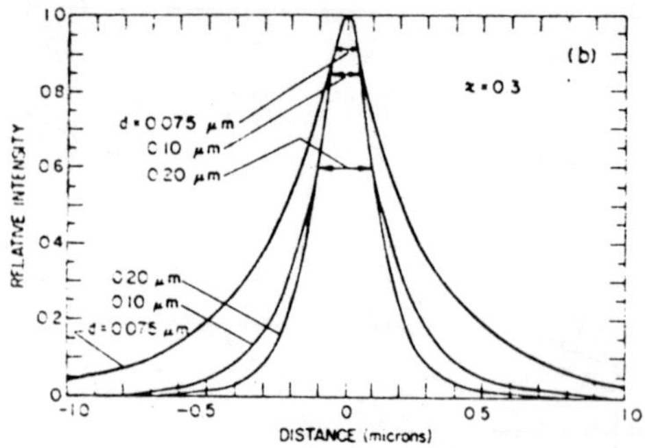

Fig. 1.5 Square of electric field as a function of position. λ=0.9 µm (Casey and Panish, 1978).

of a symmetrical waveguide is given by;

$$\Gamma_c = \left[1 + \frac{\cos^2(\kappa d/2)}{\gamma \left[d/2 + (1/\kappa)\sin(\kappa d/2)\cos(\kappa d/2) \right]} \right]^{-1} \quad (1.22)$$

1.7 Transverse Mode Control

The optical confinement in the transverse direction has been realized by introducing various types of stripe geometries to laser diodes. They can be categorized into two waveguide mechanisms: gain-guiding and index-guiding.

<u>1.7.1 Gain-guiding</u> A gain-guided laser has a narrow carrier injection region (Fig. 1.6). It was introduced to lower threshold and to stabilize transverse mode. In this structure, it is assumed that the gain profile in the active layer is expressed by g(y) as shown in the figure,

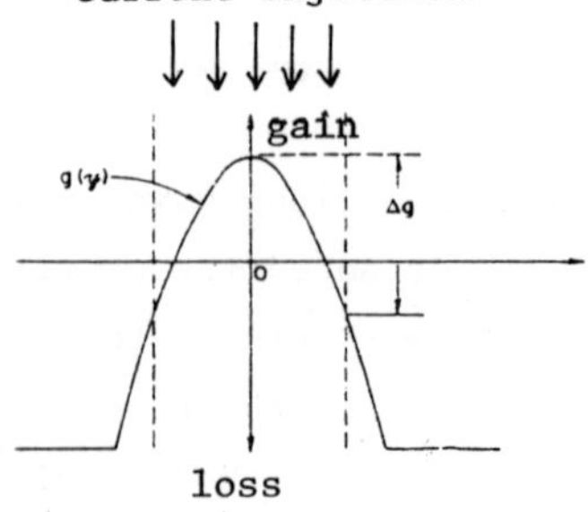

Fig. 1.6 The gain profile in a gain-guided laser.

$$g = bn(y) - c \tag{1.23}$$

where b and c are constants. The change in the refractive index due to carrier injection is

$$\Delta \bar{n}_f = -\frac{q^2 n(y)}{2m\omega^2 \varepsilon} \tag{1.24}$$

This expression means that the refractive index becomes low by current injection. Also to $\bar{n}$ is a contribution of temperature increase due to current injection $\Delta\bar{n}_T$. In total,

$$\Delta\bar{n} = \Delta\bar{n}_f \Gamma_c + \Delta\bar{n}_T \tag{1.25}$$

where Γ_c is the optical confinement factor. The magnitude $|\Delta\bar{n}|$ is on the order of 10^{-4} for injection current density of $\sim 10^{18}/cm^3$ (Paoli, 1977), and therefore may be neglected.

The electric field is then given by

$$\begin{aligned} E_y(y) &= \exp\left[-\tfrac{1}{2}(\Gamma_c/\varepsilon_0)^{1/2}(a_r + j a_i)K_0 y^2\right] \\ a_r \approx a_i &\approx (\lambda \Delta g \bar{n}(0)\varepsilon_0/\pi S^2)^{1/2} \end{aligned} \tag{1.26}$$

where the 0th mode is assumed, and ε_0 is the dielectric constant of vacuum, and S is the stripe width. The important result in the expression is the fact that the phase front of the wave is curved along the junction plane, with a virtual beam waist far behind the laser mirror. In the case of a stripe geometry laser with a 8 μm stripe width, the beam waist is 20-40 μm behind the mirror.

<u>1.7.2 Index-guiding</u> The optical intensity profile along the junction plane can also be confined to a striped region by introducing a dielectric waveguide structure, or a step refractive

index. Figure 1.7 shows a concept of such a waveguide. In the vertical direction, the wave is confined to the active layer by the double heterostructure. Since the thickness of the active layer is much thinner than the width of the stripe, it is a good approximation to replace them by a slab waveguide (Fig. 1.7(b)). The effective refractive indices $\tilde{n}_i$ are given by

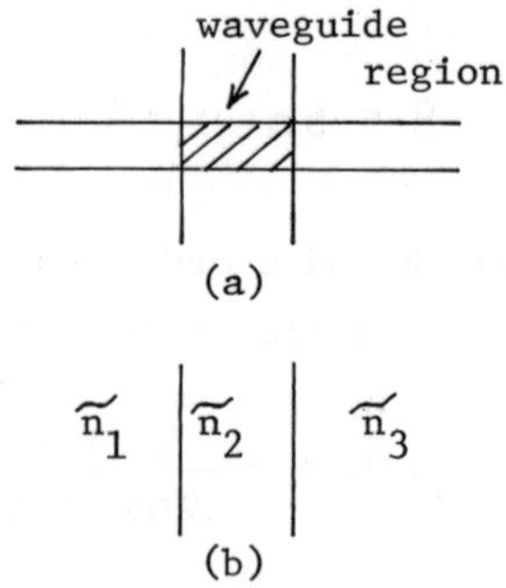

Fig. 1.7 (a) Concept of an index-guided laser. (b) Effective refractive index approximation.

$$\tilde{n}_i^2 = \int_{-\infty}^{\infty} \varepsilon \, |E_y(x)|^2 dx \Big/ \int_{-\infty}^{\infty} |E_y(x)|^2 dx \tag{1.27}$$

where $E_y(x)$ is the field intensity given by Equation 1.17. Then the electric field $E_y(x,y,z)$ is expressed by

$$E_y(x,y,z) = E_y(x)\, E_y(y) \exp(-\beta_z \cdot z) \tag{1.28}$$

where $E_y(y)$ has a form similar to Equation 1.17.

1.8 Transverse Mode Instability

1.8.1 Experimental observations When gain-guided lasers of 10-20 μm stripe widths are excited above threshold, power saturation or kinking occurs in light output-current curves (Fig. 1.8). The kink power level is typically 3-5 mW. Such lasers were found to oscillate in the fundamental transverse mode just above threshold. However, as current was increased, the mode deformed substantially, showing a kink in the light output-current curve. With a further increase in the applied current, the lasing mode deformed and centred at a different position. The displacement of the mode profile was typically 2-4 μm. The mode deformation was accompanied

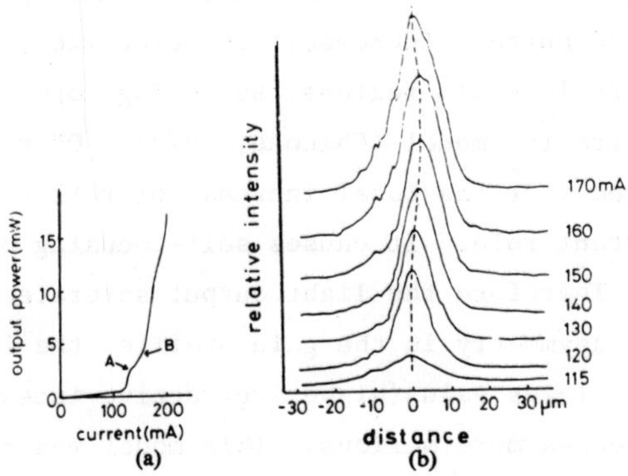

Fig. 1.8 Instability of a gain guided laser. (a)L-I curve. (b) Near-field patterns.

by a beam direction shift of 3-5°. A kink is an unstable point where two filaments are competing with each other. Thus huge optical noise is also observed at kinking. Such anomalous behavior was first thought to originate from defects or non-uniformity in the lasers. This was partially true. If a macroscopic defects exist in a cavity, stable single mode operation is hardly expected. However, the reproducibility of such instability strongly suggested that the observed instability originated from a more basic aspect in laser action.

<u>1.8.2 Models for the transverse mode instability</u> The characteristic feature on the gain-guided laser is that the waveguide mode is determined by the changeable gain profile. They change due to the local carrier consumption caused by stimulated emission of light (Lang, 1979; Kirkby et al., 1977). As the optical intensity increases, the carrier consumption becomes substantial at the peak position of lasing mode profile, and causes a dip in the carrier density profile (spatial hole burning). The local carrier consumption results in the following:

1) local decrease in gain profile,
2) local increase in refractive index profile.

By the local reduction of gain at the peak, the gain profile becomes flatter with current increase. The electric field then extends to the outer low-gain regions, suffering more loss. This is the first quantitative model (Chinone, 1977). On the other hand Kirkby et al. showed that the local increase of refractive index plays a more important role. It causes self-focusing of light to the low-gain dip. Therefore the light output saturates. If there exists a slightest asymmetry in the gain profile, the laser filament is pushed to a lower gain (higher refractive index) region, and the kinking becomes more serious. This model was theoretically investigated by Buus (1978) and Lang (1979).

For quantitative discussion, one has to simultaneously solve the wave equation and the carrier rate equation:

$$\frac{\partial^2 E_y}{\partial x^2} + (K_0^2 \bar{n}^2 - \beta^2) E = 0$$

$$\frac{\partial n}{\partial t} = D_a \frac{\partial^2 n}{\partial x^2} + \frac{J}{qd} - \frac{n}{\tau_s} - \frac{cg}{\bar{n}} |E_y|^2 \qquad (1.29)$$

where E_y : electric field amplitude

β : propagation constant

D_a : ambipolar diffusion constant

J : injection current density

τ_s : carrier life time

c : light velocity.

In the self-focusing model, the index change ratio defined by

$$R = \frac{\partial Re(\varepsilon)}{\partial n} \Big/ \frac{\partial Im(\varepsilon)}{\partial n}$$

is a key parameter. If R is negative and the absolute value is large, the self-focusing effect becomes more dominant. Numerical examples are shown in Figure 1.9. In the figure, R is assumed to be -1. It is also assumed that the dielectric constant initially has an asymmetry expressed by $\varepsilon \sim$ const + $\epsilon_A x$ with $\epsilon_A = 10^{-5}$ μm^{-1}.

It is seen in this figure that the electric field is gradually trapped into a dip in the carrier density profile. As the current is increased further, the emission spot (and the dip) moves to the higher refractive-index side.

1.9 Longitudinal Mode Control

In cleaved lasers, the wavelength of the longitudinal mode is determined by Equation 1.13. Laser action takes place at the wavelength of maximum gain. In order to control the longitudinal mode, it is necessary to introduce a mode selective waveguide structure to a laser. Distributed feedback (DFB) lasers and distributed Bragg reflector (DBR) lasers are the examples of longitudinal mode controlled lasers. They have a corrugated waveguide for optical feedback and wavelength selection.

Suppose a waveguide is corrugated as shown in Figure 1.10. The waves travelling in +z and -z directions can be described by the coupled mode equations

$$\frac{dA}{dz} = \bar{K} B \exp[-j(2\Delta\beta)z] - gA$$
$$\frac{dB}{dz} = \bar{K}^* A \exp[j\,2(\Delta\beta)z] + gB \qquad (1.30)$$

where A and B are the amplitudes of the electric fields, $\bar{K}$ is the coupling constant and $\Delta\beta = \beta - \pi \ell / \Lambda \quad (\ell = 1, 2, \cdots)$.

Assuming an input incident field of B(0) at z=0, the solutions for the forward wave E_i and the backward wave E_r are given by

$$E_i(z) = B(0)\,\frac{e^{-j\beta_0 z}\{(g - j\Delta\beta)\sinh[\bar{\gamma}(L-z)] - \bar{\gamma}\cosh[\bar{\gamma}(L-z)]}{(g - j\Delta\beta)\sinh(\bar{\gamma}L) - \bar{\gamma}\cosh(\bar{\gamma}L)}$$
$$E_r(z) = B(0)\,\frac{e^{j\beta_0 z}\sinh[\bar{\gamma}(L-z)]}{(g - j\Delta\beta)\sinh(\gamma L) - \gamma\cosh(\gamma L)}. \qquad (1.31)$$

where

$$\bar{\gamma} = (|\bar{\kappa}|^2 + (g - j\Delta\beta)^2)^{1/2}$$

The reflection gain and the transmission gain are plotted in Figure 1.11. Each plot contains four gain singularities. These are four of the longitudinal laser modes. It should be noted that the longitudinal mode closest to the Bragg frequency has the lowest threshold gain. (Yariv and Nakamura, 1977).

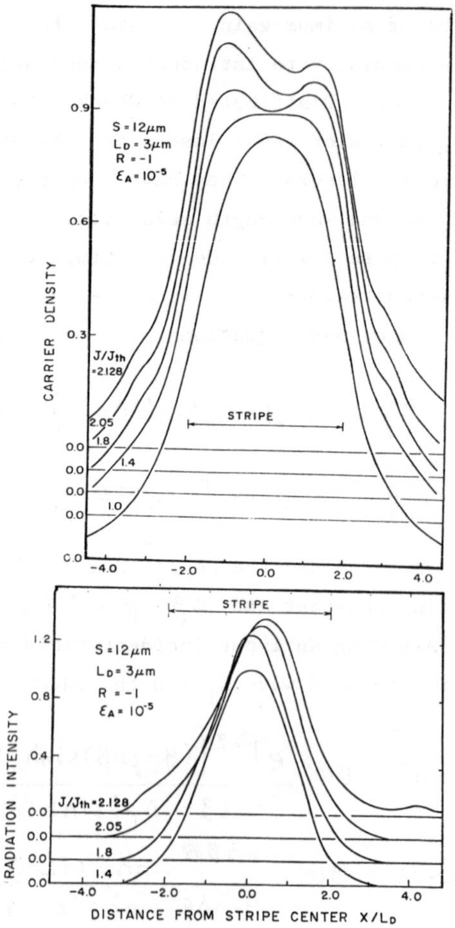

Fig. 1.9 Calculated carrier density and radiation intensity profiles of a 12 μm wide gain-guided laser (Lang, 1979).

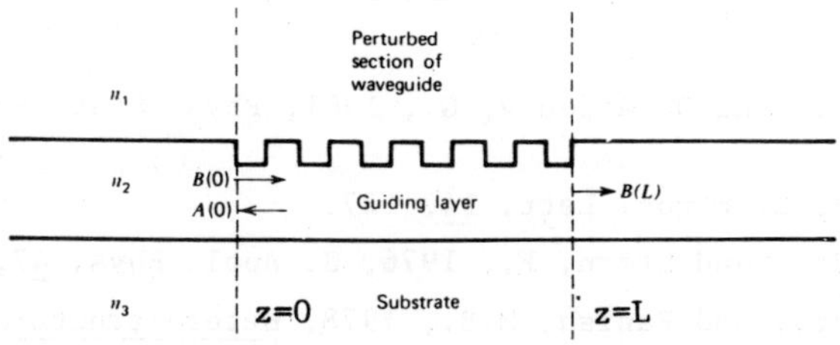

Fig. 1.10 Wave propagation in a corrugated waveguide.

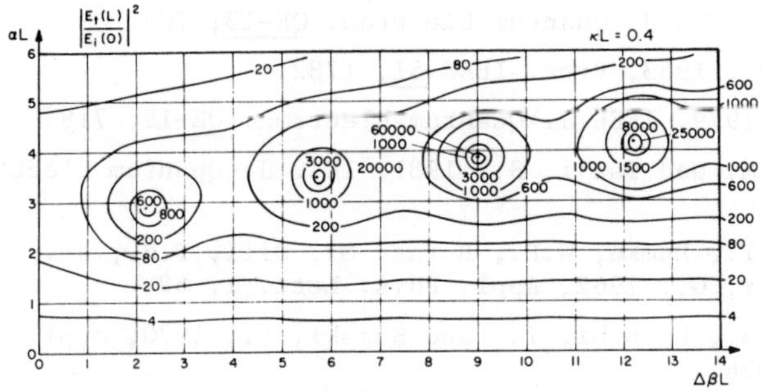

(a)

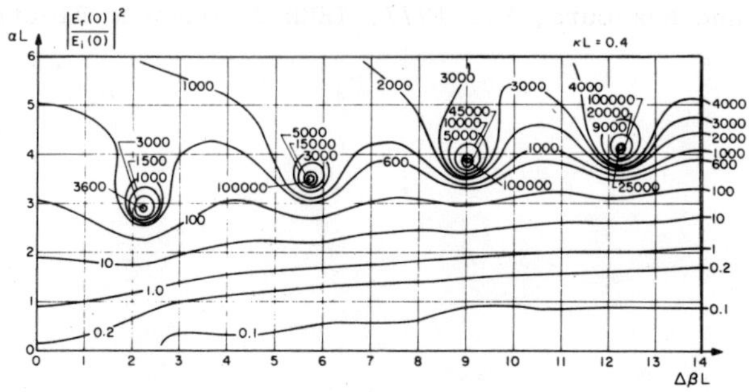

(b)

Fig. 11 The transmission gain (a) and the reflection gain (b) of a corrugated waveguide (Yariv and Nakamura, 1977).

REFERENCES

Bernard, M.G.A., and Duraffoung, G., 1961, Phys. Stat. Solidi 1, 696.

Buus, J., 1978, Electron. Lett. 14, 127.

Casey, H.C., Jr., and Stern, F., 1976, J. Appl. Phys. 47, 631.

Casey, H.C., Jr., and Panish, M.B., 1978, Heterostructure lasers (Academic Press, New York).

Chinone, N., 1977, J. Appl. Phys. 48, 3237.

Hall, R.N., Fenner, G.E., Kingsley, J.D., Soltys, T.J., and Calson, R.O., 1962, Phys. Rev. Lett. 9, 366.

Kirkby, P.A., Goodwin, A.R., Thompson, G.H.B., and Selaway, P.R., 1977, IEEE J. Quantum Electron. QE-13, 705.

Kroemer, H., 1963, Proc. IEEE 51, 1782.

Lang, R., 1979, IEEE J. Quantum Electron. QE-15, 718.

Nakamura, M., and Tsuji, S., 1981, IEEE J. Quantum Electron. QE-17, 994.

Nathan, M.I., Dumke, W.P., Burns, G., Dill, F.H., Jr., and Lasher, G., 1962, Appl. Phys. Lett. 1, 62.

Panish, M.B., Hayashi, I., and Sumski, S., 1970, Appl. Phys. Lett. 16, 326.

Paoli, T.L., 1977, IEEE J. Quantum Electron. QE-13, 662.

Quist, T.M., et al., 1962, Appl. Phys. Lett. 1, 91.

Yariv, A., and Nakamura, M., 1977, IEEE J. Quantum Electron. QE-13, 233.

2. SHORT WAVELENGTH LASERS

2.1 GaAlAs/GaAs Double Heterostructure Lasers

2.1.1 Carrier confinement The most intensively studied ternary alloy for heterostructure semiconductor lasers is GaAlAs. As is well known, the lattice constant of AlAs is very close to that of GaAs, thus GaAlAs can be grown on GaAs substrate crystals with small lattice mismatching.

The band gap energy of $Ga_{1-x}Al_xAs$, $E_g(x)$, is given by (Casey and Panish, 1978)

$$E_g(x) = \begin{cases} 1.424 + 1.247x & (\Gamma) \\ 1.90 + 0.125x + 0.143x^2 & (X) \\ 1.708 + 0.642x & (L) \end{cases} \tag{2.1}$$

The schematic band diagram of a GaAlAs/GaAs laser is shown in Figure 2.1. The carrier confinement to the active layer is achieved by the discontinuities in the conduction and valence bands, ΔE_c and ΔE_v. They are related to the electron affinities χ_1, χ_2 and the difference of the band gap energies, ΔE_g as

$$\begin{aligned} \Delta E_c &= \chi_1 - \chi_2 \\ \Delta E_v &= \Delta E_g - \Delta E_c \end{aligned} \tag{2.2}$$

For GaAlAs/GaAs lasers,

$$\begin{aligned} \Delta E_c &\sim 0.85\, \Delta E_g \\ \Delta E_v &\sim 0.15\, \Delta E_g \end{aligned} \tag{2.3}$$

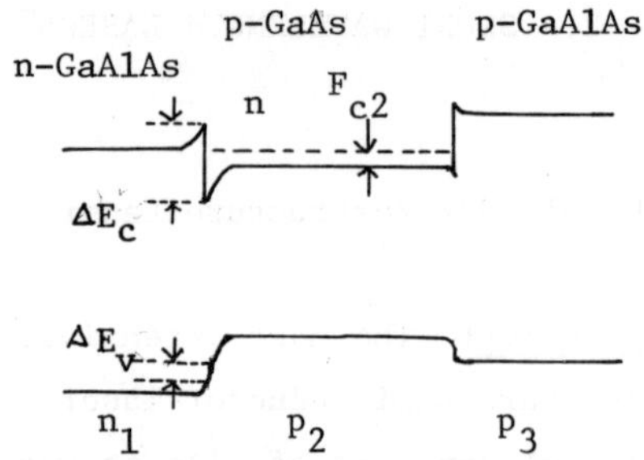

Fig. 2.1 Band diagram of a GaAs-GaAlAs DH structure.

The carrier confinement is characterized by the leakage of carriers into cladding layers. The electron leakage current i_n to the p-clad layer is obtained by

$$i_N = \frac{q D_N N_c}{L_n \tanh(W/L_n)} \left(\frac{P_2 + n}{P_3}\right) \exp\left[\frac{F_{c_2} - \Delta E_g}{kT}\right] \tag{2.4}$$

where

D_N : electron diffusion constant

N_c : effective density of state in the p-clad layer

W : thickness of the p-clad layer

P_2, P_3 : hole densities in the active and p-clad layers

n : electron density in the active layer

F_{c_2} : quesi-Fermi level of electrons in the active layer

The electron leakage current is calculated and plotted in Figure 2.2. The amount of leakage depends on the band gap energy difference, the hold concentration in the clad layer, and the electron mass in the active layer. Since the threshold current of a laser is as low as a few kA/cm^2, the leakage current should be much less than that. For GaAlAs/GaAs lasers, the AlAs mole fraction difference between the active layer and the clad layers are usually chosen to be more than 0.25 to reduce the leakage current. The hole leakage to the n-clad layer is also expressed by a similar expression. It is usually less than electron leakage current.

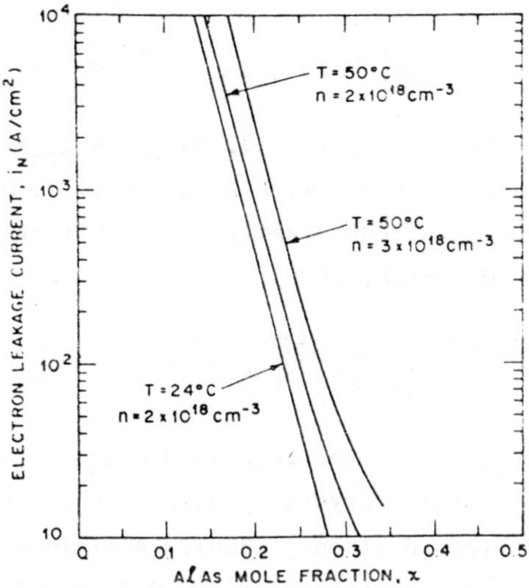

Fig. 2.2 Electron leakage current as a function of x in the p-clad layer (Casey and Panish, 1978).

2.2 Threshold current density

<u>2.2.1 Empirical expression</u> The threshold condition for laser oscillation is given by

$$g_{th} = \alpha + \frac{1}{L} \ln (1/r_1 r_2) \tag{2.5}$$

The absorption coefficient α is a sum of the free carrier absorption in the active layer (α_f), the free carrier absorption in the clad layers (α_f'), and scattering loss due to crystal defects and irregularities on interfaces (α_s). They are estimated to be

$$\begin{aligned} \alpha_f &\approx 10 - 20 \text{ cm}^{-1} \\ \alpha_f' &\lesssim 10 \text{ cm}^{-1} \\ \alpha_s &\lesssim 10 \text{ cm}^{-1} \end{aligned} \tag{2.6}$$

Therefore,

$$\alpha \approx \Gamma_c \alpha_f + (1-\Gamma_c)\alpha_f' + \alpha_s \approx 20\text{–}30 \ \text{cm}^{-1}$$

for Γ_c =0.5. For L=300 μm, $R=r_1r_2=0.3$, one gets $g_{th}=(100-120)\text{cm}^{-1}$. Using these expressions together with Equation 1.16, one can calculate the threshold current density. An empirical expression for that is (Casey and Panish, 1978)

$$J_{th}(A/cm^2) = \frac{4.5\times10^3}{\eta} d_{[\mu m]} + \frac{20 d^{[\mu m]}}{\eta \Gamma_c}\left[\alpha + \frac{1}{L}\ln(1/R)\right] \quad (2.7)$$

for d in μm, Figure 2.3 shows the threshold current density experimentally obtained by various crystal growth methods as a function of active layer thickness d (Tsang, 1980). A minimum threshold current density is usually obtained at d ~ 0.1 μm, with a value of ~ 1 kA/cm^2.

2.2.2 Wavelength dependence of threshold current Wavelength of GaAlAs lasers becomes shorter with increases in AlAs mole fraction in the active layer. Figure 2.4 shows the threshold dependence on the AlAs mole fraction or wavelength. Down to 0.75 μm, the laser threshold current density is almost constant. Below 0.75 μm, it increases due to the carrier overflow to the X and L minima. The solid line is the theoretically estimated trend of threshold (Kressel and Hawrylo, 1976).

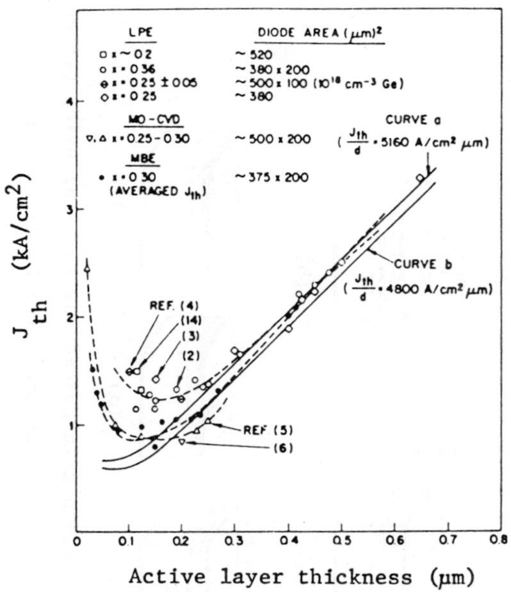

Fig. 2.3 Experimental dependence of threshold current density on active layer thickness of GaAlAs/GaAs lasers grown by LPE, MBE, and MO-CVD (Tsang, 1980).

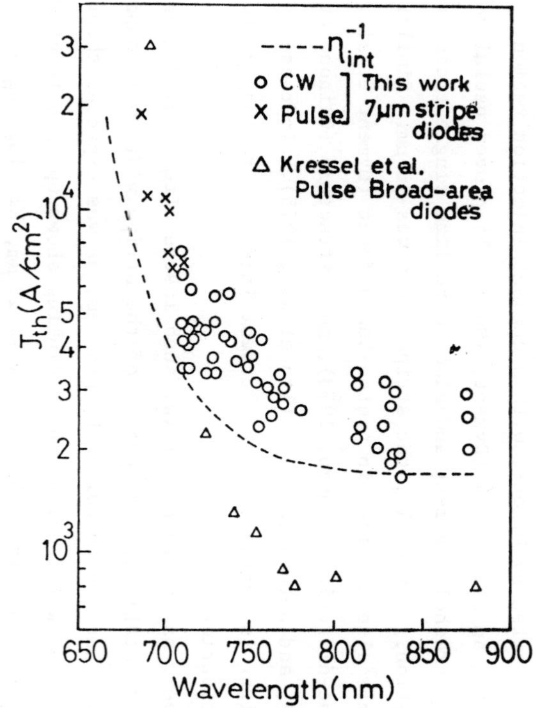

Fig. 2.4 Threshold current density of GaAlAs /GaAs lasers as a function of wavelength.

2.3 Gain-Guided Lasers

Introduction of stripe geometry into semiconductor lasers was first reported by Dyment, with a current injection region defined using an oxide mask (Dyment, 1967). It proved useful to reduce the threshold current and also to realize single transverse mode operation. Typically, the stripe width was 13 μm. Similar results were also obtained in proton-bombardment (Dyment et al., 1972), zinc-diffusion (Yonezu, 1973), mesa-structure (Chinone et al., 1976), and heteroisolation (Ito et al., 1975). In these lasers, light is confined to the current injection region due to gain distribution (Fig. 2.5).

However, as mentioned before, such lasers often showed transverse mode instability. Reduction of the stripe width was found to be most effective to solve the problem, as was expected theoretically. The power level at which kinking shows up is a function of stripe width w (Fig. 2.6). For $w \sim 13$ μm, kinking appears at $P \sim 3$ mW. For $w \sim 6$ μm, the transverse mode is stable up to $P \sim 5$ mW (Kobayashi et al., 1977). A recent example of a gain-guided laser is a V-grooved laser. The current is limited to a V groove, resulting in a better carrier confinement (Marschall et

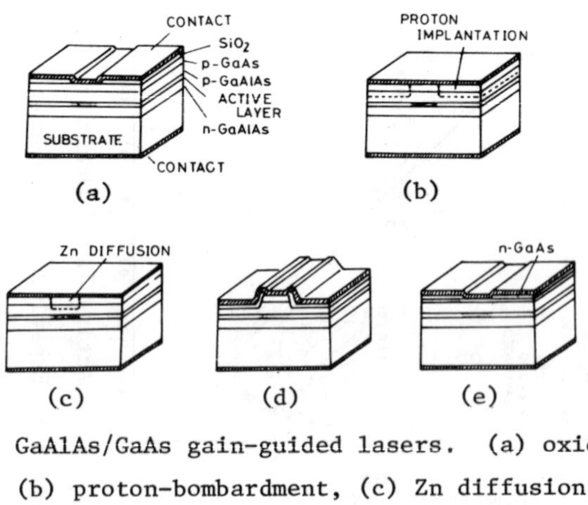

Fig. 2.5 GaAlAs/GaAs gain-guided lasers. (a) oxide stripe, (b) proton-bombardment, (c) Zn diffusion, (d) mesa-structure, (e) heteroisolation.

al., 1978). The room temperature threshold was 50 mA.

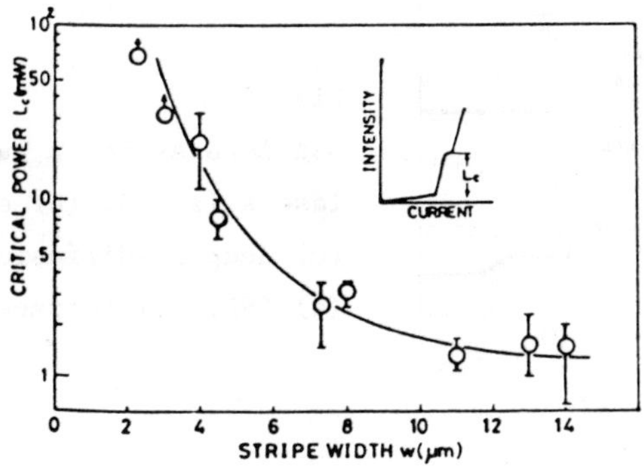

Fig. 2.6 Critical power of narrow stripe GaAlAs/GaAs lasers (Kobayashi et al., 1977).

2.4 Index-Guided Lasers

Numerous structures of rigid index-guided lasers have been demonstrated (Fig. 2.7). The BH laser has a filamentary GaAlAs active region embedded in GaAlAs clad with a larger AlAs mole fraction (Tsukada, 1974). The AlAs mole fraction is chosen to be ~ 0.05 for the active layer and ~ 0.4 for the cladding GaAlAs. Such a change in the mole fraction introduces an index step of ~ 0.25 in the vertical and horizontal directions. The stripe width is typically 1-2 μm. In order to have higher optical power, a large optical cavity structure was also introduced to the BH structure (Saito and Ito, 1980). In this case, cw output power of ~ 10 mW is obtained without facet degradation. The threshold current density of the BH laser is 20-30 mA. The interesting feature of the large optical cavity BH laser is its high quantum efficiency. It is due to the perfect carrier confinement to the active region and the smaller optical absorption in the low carrier density waveguide layer. An external quantum efficiency of 80% was obtained.

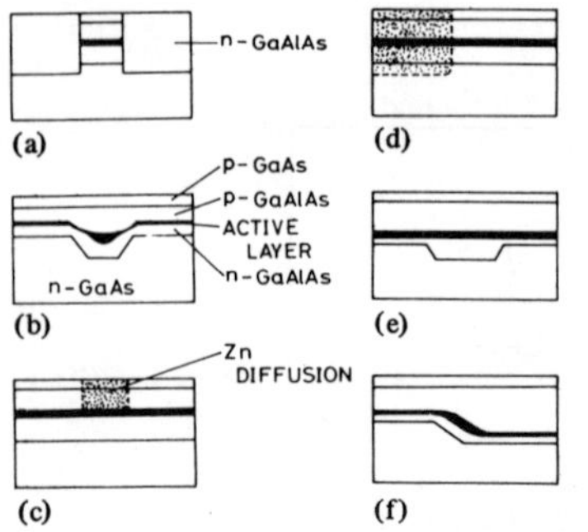

Fig. 2.7
GaAlAs/GaAs index guided lasers (a) BH, (b) embedded, (c) deep Zn-diffusion, (d) TJS, (e) CSP, (f) terraced substrate.

The laser structure similar to the BH structure is the embedded structure (Fig. 2.7 (b)), where a curved active layer is embedded in GaAlAs. Threshold current as low as 30 mW and single stable mode operation up to 5 mW were reported (Kirkby et at., 1976; Burnham et al., 1975). An index step can be obtained by a change in doping level due to the free carrier effect. The examples of such lasers are deep zinc-diffusion lasers (Yonezu et al., 1977) and TJS lasers (Namizaki, 1975). In TJS lasers, the active layer is Te-doped with a carrier concentration of 2×10^{18} cm^{-3}. Zinc is selectively diffused by a two step diffusion method. As a result, carrier density has a dip around the p-n junction in the active layer, resulting in optical confinement. Linear light output-current curves were obtained up to 20 mW with threshold of 20-30 mA. In the deep zinc-diffused laser, the zinc is diffused to the active layer with a diffusion width of 4-8 μm. The change in conduction type results in positive index step in the direction parallel to the junction.

The third method to realize stable transverse mode operation is to grow double heterostructure on a grooved substrate. A channeled substrate planar structure is an example (Aiki et al., 1977) where a planar double heterostructure is formed on a 4 μm wide and 1 μm deep groove. The thickness of the first n-GaAlAs clad layer is ~ 0.4 μm outside the groove. Due to the change in the effective index in the horizontal direction, the light is well confined to the active region just above the groove. The threshold is as low as 70 mA and cw output of 15 mW is obtained without facet degradation. Another example of laser structures formed on grooved substrates are plano-convex waveguide (PCW) structure (Ide et al. 1980), constricted double heterostructure (CDH) (Botez, 1978), and a terraced substrate structure (TS) (Sugi, 1979). In these structures, the thickness of the active layer varies lens-like, resulting in transverse mode control.

2.5 Longitudinal Mode Behavior

Semiconductor lasers show a variety of spectral behaviors, depending on laser structure, mode stability, etc. In semiconductor lasers with cleaved facets, the spectral peak shifts with applied current and temperature. The number of longitudinal modes was found to depend strongly on waveguide mechanism.

2.5.1 Lasing spectra of a gain-guided laser It is well known that gain-guided lasers with a narrow stripe width ($w \sim 5\ \mu m$) lase in multi-longitudinal modes. The spectrum of such a laser is shown in Figure 2.8. The number of the lasing longitudinal modes is typically 10-15 at $I \sim 1.5\ I_{th}$. The large contribution of spontaneous emission of light is the origin of the multimode operation (Peterman, 1979). By assuming the spontaneous emission factor $\beta_s \sim 10^{-4}$, one can explain the multimode behavior.

2.5.2 Spectra of an index-guided laser The longitudinal mode behavior of index-guided lasers is less complicated. It is semi-quantitatively understood by the rate equation, using β_s of 10^{-6}. In this case, the spectrum is virtually single at excitation levels above 1.1 times threshold. Such spectral narrowing is the evidence of the homogeneous broadening character of the semiconductors. Figure 2.8 (b) shows the spectra of a channeled substrate planar structure laser under cw operation, where the

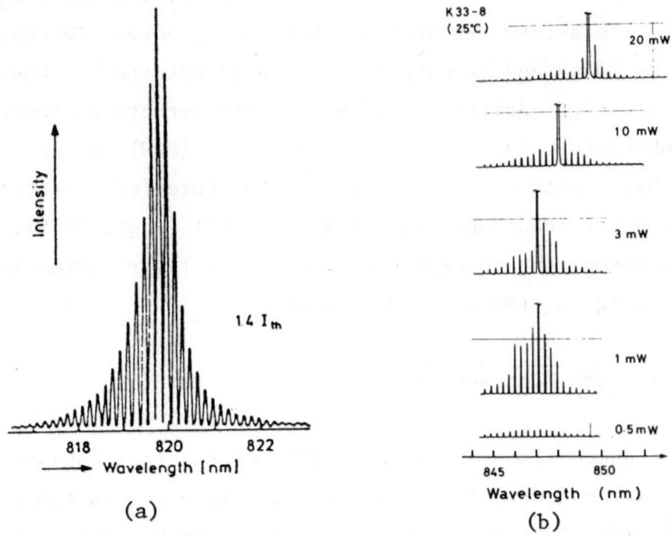

Fig. 2.8 Lasing spectra of GaAlAs/GaAs lasers. (a)gain-guided laser, (b)index guided laser.

intensity scales are the same for the five spectra shown. Most of the lasing power is seen to concentrate in a single longitudinal mode whose intensity is off the scale at output powers of 1 mW or more.

An interesting finding on the longitudinal mode behavior is the hysteresis observed in the lasing wavelength-temperature curves or in the lasing wavelength-current curves. It is a direct indication of strong coupling in semiconductor lasers (Nakamura et al., 1978).

2.5.3 Linewidth The linewidth, i.e. the coherence of single mode laser light, was investigated by using a carefully arranged Fabry-Perot etalon system (Takakura et al., 1980), and a delayed self-heterodyne method (Ohkoshi et al., 1980).

The half spectral width is theoretically given by

$$\delta\nu = \frac{2\pi h\nu\,(\Delta\nu_c)^2}{P}$$

(2.8)

where $h\nu$ is the laser photon energy, $\Delta\nu_c$ is the half line width of the resonator, and P is the output power. For a channeled substrate planar structure laser, a line width of 1 MHz was obtained, which is consistent with the above equation with $\Delta\nu_c \sim$ 26 GHz. For a BH laser emitting at 0.84 μm, $\Delta\nu_c$ is 75 GHz. The stabilization of lasing frequency was also studied. A frequency drift of less than 10 MHz was obtained in 30 minutes under temperature stabilized operation.

<u>2.5.4 Mode partition noise</u> When a laser is oscillating in multi-longitudinal modes, the intensity of each mode is fluctuating significantly. The fluctuation is more than 30 dB larger than the total power fluctuation. It is due to the random process of electron-hole recombination in the laser. If the external optical circuit has a frequency selectivity function, it will cause intensity fluctuation of the laser light.

When a few modes are competing with each other, in particular, the total output power itself is also observed to change from time to time depending which mode is more dominant.

2.8 Longitudinal Mode Control by Grating Feedback

The GaAlAs/GaAs DFB lasers were fabricated and cw operated at room temperature. The grating period was determined by

$$\Lambda = \lambda l / 2\bar{n}_g \tag{2.9}$$

where $\bar{n}_g$ is the effective index of the waveguide, and l is an integer. Experiments were mostly done for $l = 3$ (third order grating), with Λ = 0.34-0.36 μm. Figure 2.9 shows the threshold current density and wavelength of a DFB laser as a function of temperature (Nakamura et al., 1975). For comparison, the corresponding data of a Fabry-Perot laser of the same wafer are also shown. DFB lasers operated in a single longitudinal mode by the mode selectivity of the grating. The temperature dependence of the lasing wavelength was 0.8 Å/deg, which was due to the temperature dependence of the refractive index. The threshold current density was ~ 3 kA/cm^2 at room temperature. The threshold is a little bit high partly because the grating was separated from the active layer to minimize the effect of crystal damages caused by the grating fabrication.

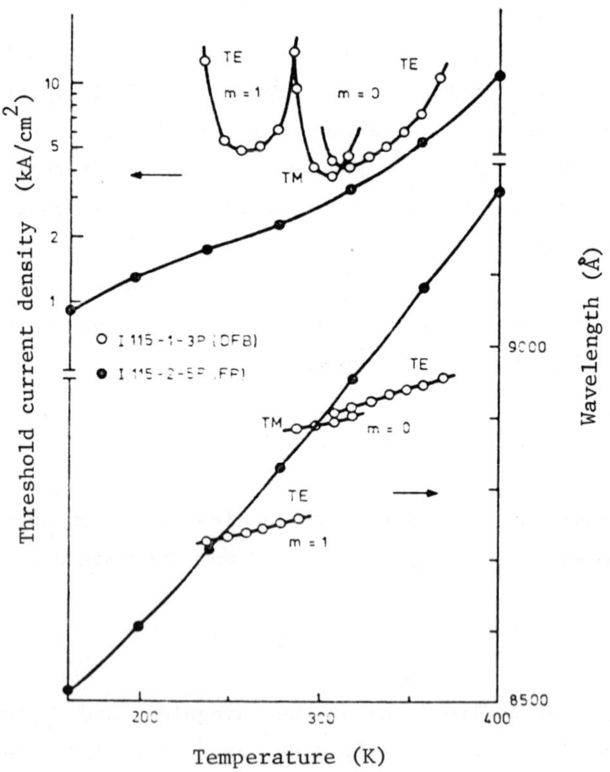

Fig. 2.9 Threshold current density and lasing wavelength of a GaAlAs/GaAs DFB laser (Nakamura et al., 1975).

REFERENCES

Aiki, K., Nakamura, M., Kuroda, T., and Umeda, J., 1977, Appl. Phys. Lett. 48, 649.

Botez, D., and Zory, P., 1978, Appl. Phys. Lett. 32, 261.

Burnham, R.D., and Scifres, D.R., 1975, Appl. Phys. Lett. 27, 510.

Casey, H.C., Jr., and Panish, M.B., 1978, Heterostructure Lasers (Academic Press).

Chinone, N., Ito, R., and Nakada, O., 1976, J. Appl. Phys. 47, 785.

Dyment, J.C., 1967, Appl. Phys. Lett. 10, 84.

Dyment, J.C., et al., 1972, Proc. IEEE 60, 726.

Ide, Y., et al., 1980, Appl. Phys. Lett. 36, 121.

Ito, K., Inoue, M., and Teramoto, I., 1975, IEEE J. Quantum Electron., QE-11, 421.

Kirkby, P.A., and Thompson, C.H.B., 1976, J. Appl. Phys. 47, 4578.

Kobayashi, T., Kawaguchi, H., and Furukawa, Y., 1977, Japan J. Appl. Phys. 16, 601.

Kressel, H., and Hawrylo, F.Z., 1976, Appl. Phys. Lett. 28, 598.

Marschall, P., Schlosser, E., and Wolk, C., 1978, in Proc. 4th European Conf. Opto. Commun., 94.

Nakamura, M., et al., 1975, IEEE J. Quantum Electron. QE-11, 436.

Nakamura, M., et al., 1978, J. Appl. Phys. 49, 4644.

Namizaki, H., 1975, IEEE J. Quantum Electron. QE-11, 427.

Ohkoshi, T., Kikuchi, K., and Nakayama, A., 1980, Electron. Lett. 16, 630.

Peterman, K., 1979, IEEE J. Quantum Electron. QE-15, 566.

Saito, K., and Ito, R., 1980, IEEE J. Quantum Electron. QE-16, 205.

Sugi, T., et al., 1979, Appl. Phys. Lett. 34, 270.

Takakura, T., Tako, T., and Iga, K., 1978, private communications.

Tsang, W.T., 1980, Appl. Phys. Lett. 36, 11.

Tsukada, T., 1974, J. Appl. Phys. 45, 4899.

Yonezu, H., et al., 1973, Japan J. Appl. Phys. 12, 1285.

Yonezu, H., et al., 1977, Japan J. Appl. Phys. 16, 209.

3. LONG WAVELENGTH LASERS

3.1 InGaAsP/InP Double Heterostructure Lasers

InGaAsP/InP lasers are of particular interest because they operate in the wavelength region where the fiber absorption loss becomes minimum (0.3-0.5 dB/km). The energy gap-lattice constant-composition diagram of the quaternary material is shown in Figure 3.1 (Moon et al., 1974). The broken line from the upper left-hand corner represents the lattice matched compositions to InP, corresponding to photon energy of 0.74 to 1.35 eV (1.7-0.9 μm). The band gap energy and the refractive index of $In_{1-x}Ga_xAs_yP_{1-y}$ lattice matched to InP are given by (Moon et al., 1974; Nahory et al., 1978)

$$E_g(y) = 1.35 - 0.72y + 0.12y^2$$
$$\bar{n}(y) = 3.4 + 0.256y - 0.095y^2 \tag{3.1}$$

where the lattice matched condition is

$$y = 2.21x / (1 + 0.686x) \tag{3.2}$$

The emission photon energy is about 0.03 eV less than the band gap energy. For 1.3 and 1.55 μm, the active layers are $In_{0.72}Ga_{0.28}As_{0.59}P_{0.41}$ and $In_{0.58}Ga_{0.42}As_{0.90}P_{0.10}$ respectively.

Since the first operation of InGaAsP/InP lasers, the double heterostructure has been fabricated by liquid phase epitaxy at 600-650°C. The lattice mismatch is controlled to within 0.03% by controlling the melt composition and growth temperature. The thermal expansion coefficients of InP and InGaAsP are $4.6 \times 10^{-6}/°C$ and $5.4 \times 10^{-6}/°C$, respectively (Bisaro et al., 1979). If the lattices are matched at the growth temperature, mismatching is about 0.06% at room temperature.

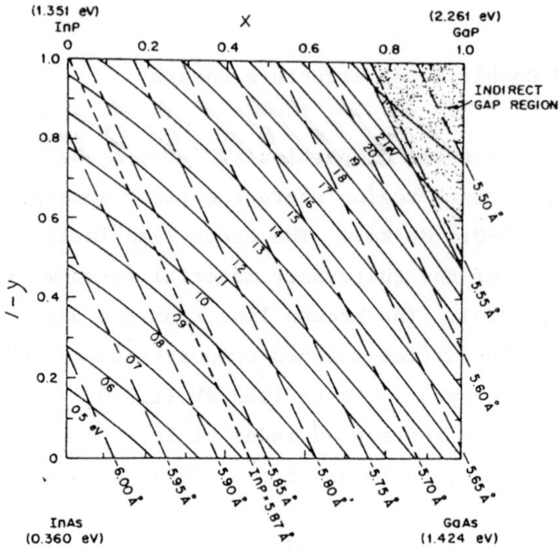

Fig. 3.1 The x-y compositional plane for $In_{1-x}Ga_xAs_yP_{1-y}$ at 300 K (Moon et al., 1974).

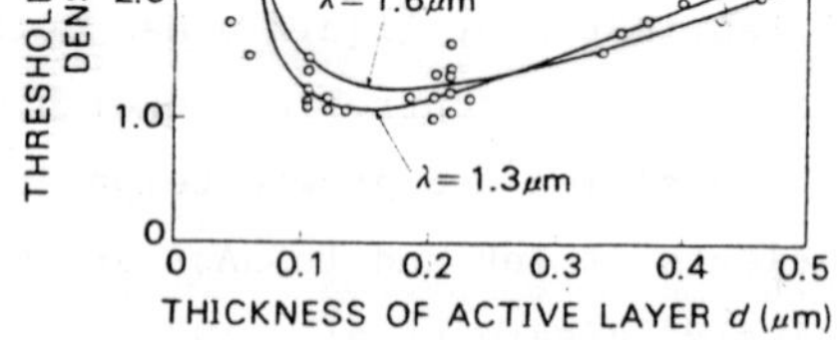

Fig. 3.2 Threshold current density of 1.3 μm lasers as a function of active layer thickness (Arai et al., 1982).

3.2 Threshold Current Density

The threshold current density of InGaAsP/InP lasers is shown as a function of the thickness of the active layer in Figure 3.2. Threshold currents of ~ 1 kA/cm^2 are obtained at d ~ 0.15 μm, which is comparable to those of GaAlAs/GaAs lasers (Arai et al. 1982).

The threshold current density of the InGaAsP/InP laser is almost consant in the wavelength region of 1.2-1.6 μm. At wavelengths below 1.2 μm, the threshold current density increases substantially due to the decrease of the barrier height at the heterojunction (ΔE_g = 0.23 eV at λ = 1.1 μm).

3.3 Temperature dependence of threshold current

The threshold current density of semiconductor lasers is approximately expressed by

$$J_{th} = J_0^* \exp (T/T_0) \tag{3.3}$$

where J_0^* and T_0 are constants. The temperature dependences of GaAlAs/GaAs lasers and InGaAsP/InP lasers are shown in Figure 3.3. In the case of GaAlAs/GaAs lasers, the characteristic temperature T_0 is $\gtrsim$ 150 K, resulting from the temperature dependence of peak gain in the GaAlAs active region. In the case of InGaAsP/InP lasers of 1.2-1.6 μm, the characteristic temperature is typically 60-70 K at room temperature. From the peak gain dependence on wavelength, one could expect T_0 of 150 K or higher. The characteristic temperature was also found to be almost independent of lasing wavelength for λ = 1.2-1.6 μm.

In order to explain the discrepancy, the following mechanism have been proposed.

<u>3.3.1 Leakage current over heterojunctions</u> Assuming that the electron effective mass m_e is $\sim 0.04\ m_0$ (m_0 = free electron mass) for ~ 1.3 μm, Yano et al. obtained a theoretical value of T_0 =

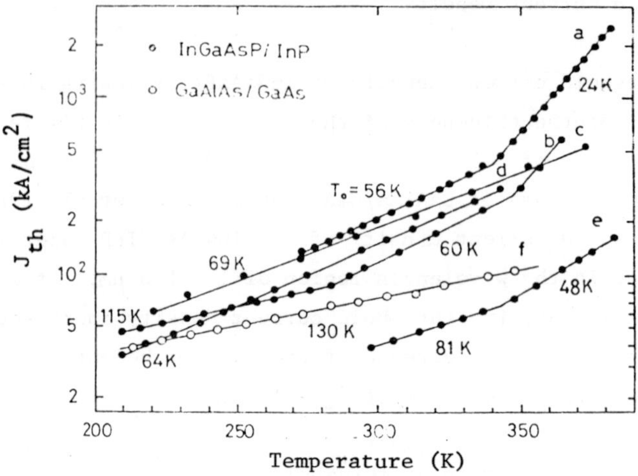

Fig. 3.3 Temperature dependence of threshold current density.

60–110 K (Yano et al., 1980). Spontaneous emission due to the carrier leakage into the p-InP clad layer was also observed.

<u>3.3.2 Nonradiative recombination at heterointerfaces</u> Nonradiative recombination of carriers seems to take place in the InGaAsP/InP laser, making carrier lifetimes shorter at higher temperatures. However, the interface recombination model has been rejected as a major mechanism since multilayered InGaAsP/InP lasers were shown to have similar temperature characteristics (Rezek et al., 1980).

<u>3.3.3 Nonradiative recombination through Auger process</u> The most plausible mechanism is the Auger recombination of carriers. The band to band Auger processes were investigated theoretically and found to explain the observed temperature characteristics well (Sugimura, 1981; Dutta and Nelson, 1982). Figure 3.4 shows three types of band to band Auger processes. In these processes, the injected electrons recombine with holes by exciting other electrons in conduction bands (CCCH), holes from the spin split-off band to the heavy-hold valence band (CHHS), and holes from the light-hole valence band to the heavy-hole valence band (CHHL). The CCCH

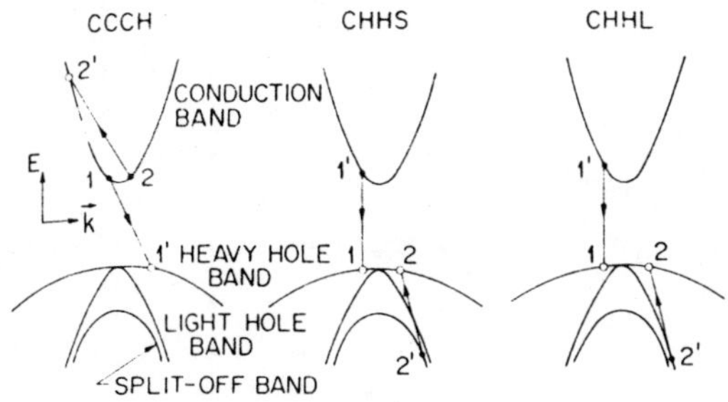

Fig. 3.4 Three band to band Auger processes. Electrons are represented by closed circles. Holes are represented by open circles (Dutta and Nelson, 1982).

process is important for n-type materials, and CHHS and CHHL processes for p-type materials. At high injection current, all of them could be important. Figure 3.5 shows calculated Auger life time. The CHHS rate is about one order of magnitude higher than the rate of CHHL process. One notes that the Auger lifetime decreases rapidly with increasing temperature. The calculated T_0 for the 1.3 μm InGaAsP/InP laser is about 50 K.

Other Auger processes such as photon-assisted ones were also studied theoretically and found to be less significant to explain the small value of T_0.

The auger processes were also used to explain the power saturation of InGaAsP/InP LEDs at higher current levels. To do so, an n^3 term is added to the rate equation of injected carriers.

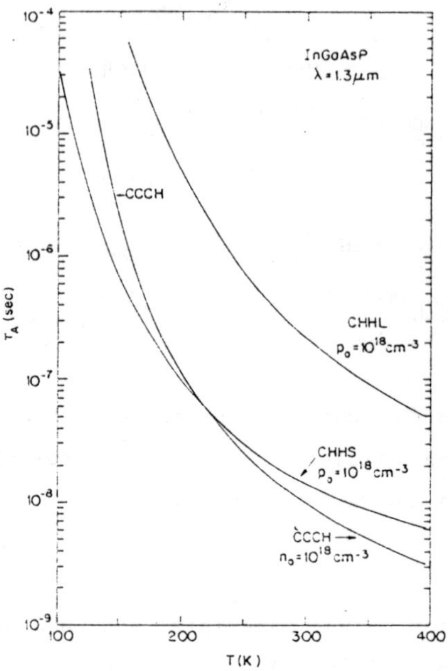

Fig. 3.5 Auger lifetime for InGaAsP ($\lambda = 1.3$ μm) (Dutta and Nelson, 1982).

<u>3.3.4 Laser light absorption by valence electrons</u> Light absorpion due to the transition between the split-off and heavy-hole valence bands were calculated by Asada et al. (1981) and concluded to explain the temperature characteristics of the laser threshold.

It is still difficult to say quantitatively how much each process contributes to degrade the temperature characteristics. It also depends on device parameters such as the active layer thickness and the doping levels.

3.4 Gain-Guided Lasers

Various types of stripe geometry lasers were demonstrated in InGaAsP/InP lasers. But because of their high threshold current ($I_{th} \sim 150$ mA), it was difficult to operate them above 50°C under cw conditions. The reduction of threshold current is especially important because of the strong temperature dependence of threshold current. In this sense, no satisfactory gain-guided lasers have been demonstrated in the wavelength region of 1.1-1.6 μm.

3.5 Index-Guided Lasers

The introduction of index steps in the horizontal direction has turned out to be useful again in InGaAlAs/InP lasers. In Figure 3.6 are summarized typical waveguide structures. The bonus of such laser structures is a better carrier confinement, resulting in the reduction of threshold current (less than 70 mA) and the capability of high temperature operation.

Among the successful structures, the BH structure has the lowest threshold of 20-30 mA at room temperature. The structure is fabricated by growing an InGaAsP/InP double heterostructure on an InP substrate, and then forming a mesa structure by chemical etching, and finally growing InP layers in the etched portions.

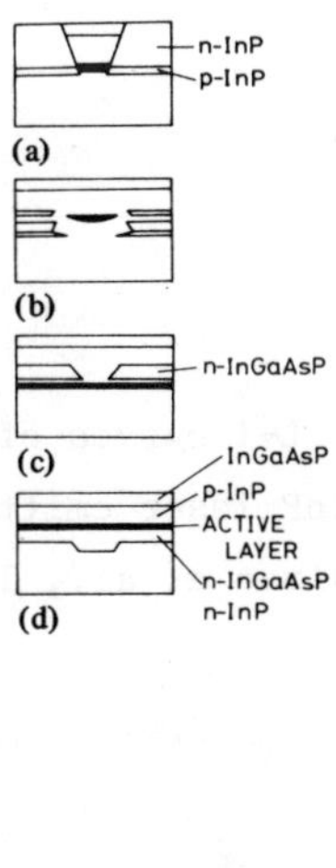

Fig. 3.6 Index-guided lasers of InGaAsP/InP. (a) BH, (b) crescent, (c) self-aligned, (d) PCW.

The lattice constant of the active layer is determined by x-ray diffraction experiments and controlled to be within ±0.03% from the lattice constant of InP. The mesa structure is formed by Br-methanol solution using SiO_2 stripes as masks. The Br-methanol solution, because of its slow etching speed for the (111) InP surface, makes it possible to form a narrow active region parallel to the 〈 110 〉 direction.

The room temperature light output-current curves of a BH laser emitting at 1.3 μm are depicted in Figure 3.7. The laser threshold is 22 mA in pulsed operation and 24 mA for cw operation. A differential quantum efficiency of 60% for pulsed operation, which is comparable with that of GaAlAs/GaAs lasers, implies the high quality of the crystals (Hirao et al. 1980).

A similar narrow stripe laser with a crescent-shaped active region has been developed using two step liquid phase epitaxy (Murotani et al., 1980). In the fabrication, an n-InGaAsP layer, p-InGaAsP layer, an n-InGaAsP layer were sequentially grown on an InP substrate. Then a channel with a width of 4 μm and a depth of 2 μm was formed by chemical etching. Finally the double heterostructure was formed on the grooved substrate. The active region has a crescent form embedded in InP. A threshold current as low as 28 mA was reported.

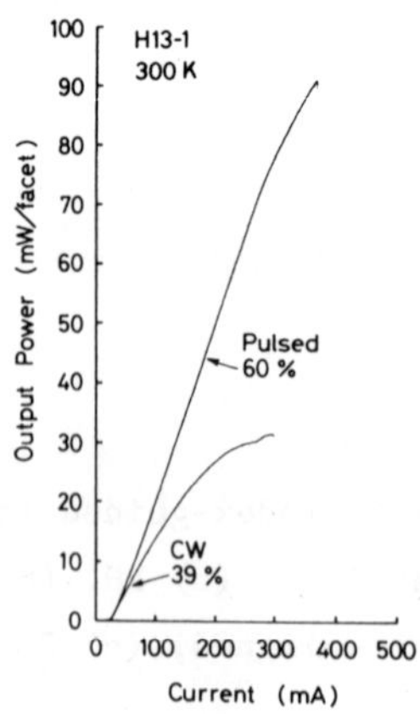

Fig. 3.7 L-I curves of a BH InGaAsP/InP laser emitting at 1.3 μm (Hirao et al., 1980).

Another example of a transverse mode controlled laser is a self-sligned structure (SAS) (Fig. 3.6). In this laser, light absorbing n-InGaAsP regions are formed above the InGaAsP active layer. Therefore the waveguide mechanism is similar to that of a CSP laser of GaAlAs/GaAs (Yano et al., 1979). In the SAS laser, the light absorbing regions also work to confine the injection current to the lasing region. The window width is 5 μm. Room temperature threshold current was reported to be ~ 70 mA.

The plano-convex waveguide (PCW) structure has an n-InGaAsP active layer on a grooved InP substrate (Ueno et al., 1981). The width and the depth of the groove are 4 and 0.4 μm, respectively. Because of the thickness change in the waveguide layer, an effective index change of 0.05-0.08 is realized in the transverse direction, which stabilizes the laser transverse mode. The room temperature threshold current is ~ 70 mA.

3.6 Temperature Limit for CW Operation

The maximum temperature for cw operation is estimated by solving simultaneously the following equations;

$$I_{th}^{cw} = I_0^* \exp(T + \Delta T / T_0) \qquad (I_0^*: \text{const})$$
$$\Delta T = (V_j + I_{th}^{cw} R_s)\, I_{th}^{cw} R_{th} \tag{3.4}$$

where

I_{th} : threshold current for cw operation
V_j : diffusion voltage
R_s : series resistance
R_{th} : effective thermal impedance

Figure 3.8 illustrates the maximum heatsink temperature for cw operation of the InGaAsP/InP BH lasers with 1-2 μm stripe widths as a function of pulsed threshold current at 20°C. The solid and the dashed lines are numerical results obtained from the above equations.

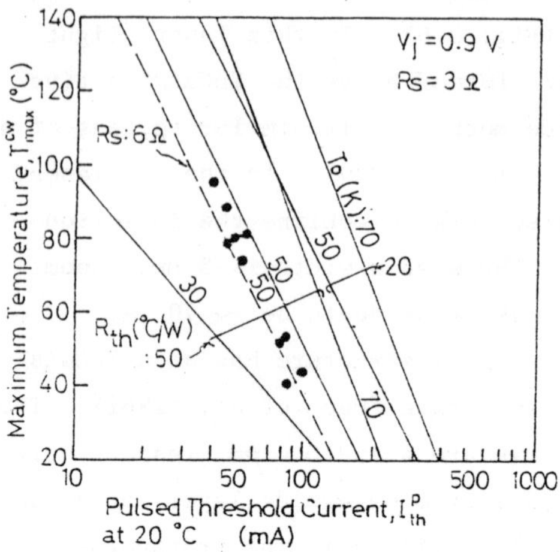

Fig. 3.8 Maximun heatsink temperature for cw operation of BH laser.

It is noted that for a given pair of R_{th} and T_0, the reduction of threshold current is crucial for high temperature operation of the lasers.

For the room temperature threshold of ~ 30 mA, the maximum temperature is as high as 100°C with T_0 of 50-70 K, R_s of 3-6 ohms, V_j of 0.9 V, and R_{th} of 40-50°C/W in the BH lasers. The highest temperature obtained is 110°C. For the lower threshold region (~ 10 mA), the maximum operation temperature does not become as high as predicted in Equation 3.4. It is because the leakage current through the buried regions becomes predominant at high temperatures in such a low threshold narrow stripe laser.

3.7 Longitudinal Mode Control by Grating Feedback

The study on the InGaAsP/InP DFB and DBR lasers is growing recently. It is because the rather high material dispersion at 1.55 μm, where the fiber loss becomes minimum, requires narrow spectrum for high bit rate long distance optical transmission. In the case of conventional cleaved lasers, even the index-guided lasers operate in multilongitudinal mode under high bit rate direct modulation because of the overshoot of gain above the threshold gain of laser operation. The wavelength selectivity by grating feedback solves this problem.

3.7.1 DFB lasers operating at 1.5 μm A schematic structure is shown in Figure 3.9 (Matsuoka et al., 1981). It has a grating of 0.45 μm period and 0.2 μm depth fabricated directly on the (100) InP substrate. The grating lines are aligned to the ⟨110⟩ direction. On the corrugated surface of the substrate are grown an n-InGaAsP waveguide layer, a p-InGaAsP active layer, a p-InP clad layer, and a p-InGaAsP cap layer. The waveguide thus has a LOC structure. Using such a crystal, a BH structure is formed by the conventional method for transverse mode control and carrier confinement. A similar structure was also demonstrated by Akiba et al (Akiba et al., 1981), where a buffer layer was formed between the active layer and the p-clad layer to prevent crystal meltback during liquid phase epitaxy.

The threshold current densities and lasing wavelengths of a DFB BH laser and a Fabry-Perot BH laser are shown in Figure 3.10. The DFB laser operated in a TE mode with a temperature sensitivity of 1.0 Å/deg. Such lasers were cw operated up to 310 K in the 1.5 μm region. High frequency experiments were done at 400 MHz, I_{p-p} = 20 mA, confirming the single mode operation at high frequencies.

3.7.2 DBR lasers operating at 1.5 μm Similar experiments were performed with DBR lasers. A stable single mode operation was obtained for modulation frequencies ranging from 0.25 GHz to 3 GHz at 100% modulation depth (Koyama et al., 1981).

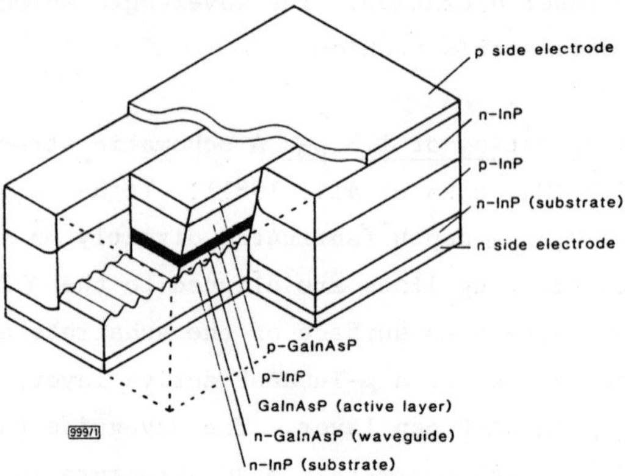

Fig. 3.9 Schematic diagram of DFB-BH InGaAsP/InP laser (Matsuoka et al., 1982).

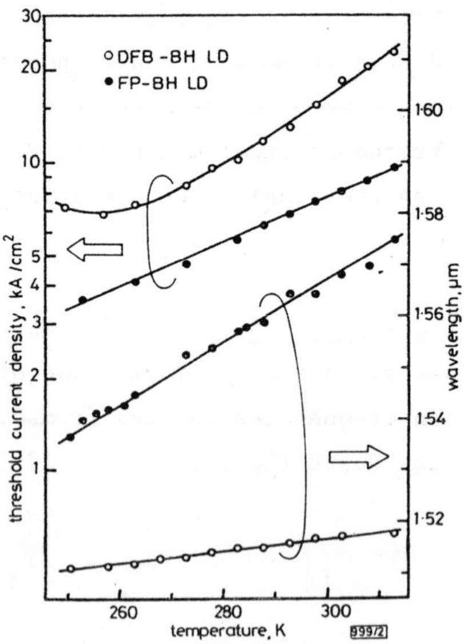

Fig. 3.10 Threshold current density and lasing wavelength of a DFB-BH InGaAsP/InP laser (Matsuoka et al., 1982).

REFERENCES

Akiba, S., et al., 1981, Electron. Lett., 18, 78.

Arai, S., et al., 1982, in Optical Devices and Fibers, edited by Y. Suematsu (OHM, Tokyo).

Asada, M., et al., 1981, IEEE J. Quantum Electron. QE-17, 611.

Bisaro, R., Merenda, P., and Pearsall, T.P., 1978, Appl. Phys. Lett., 34, 100.

Dutta, N.K., and Nelson, R.J., 1982, J. Appl. Phys. 53, 74.

Hirao, M., et al., 1980, J. Appl. Phys. 51, 4539.

Koyama, F., et al., 1981, Electron. Lett. 18, 28.

Matsuoka, T., et al., 1981, Electron. Lett. 18, 28.

Moon, R.L., Antypas, G.A., and James, L.W., 1974, J. Electron. Materials 3, 635.

Murotani, T., et al., 1980, Electron. Lett., 16, 566.

Nahory, R.E., et al., 1978, Appl. Phys. Lett. 33, 659.

Rezek, E.A., Holonyak, N., Jr., and Fuller, B.K., 1980, J. Appl. Phys. 51, 2404.

Sugimura, A., 1981, IEEE J. Quantum Electron. QE-17, 627.

Yano, M., Nishi, H., and Takusagawa, M., 1979, IEEE J. Quantum Electron. QE-15, 1388.

Yaon, M., Nishi, H., and Takasagawa, M., 1980, J. Appl. Phys. 51, 4020.

Ueno, M., et al., 1981, IEEE J. Quantum Electron. QE-17, 1930.

4. RELIABILITY OF SEMICONDUCTOR LIGHT SOURCES

The reliability of semiconductor light sources has been of great concern. Many origins have been assigned as their degradation mechanisms. Among them are DLD growth, facet oxidation, and catastrophic facet melting.

4.1 Degradation Mechanisms of GaAlAs/GaAs Lasers

The degradation of GaAlAs/GaAs lasers may be conveniently separated into three categories:

(1) dark line defect (DLD) formation,

(2) mirror damage, and

(3) gradual degradation.

These are closely related to fabrication process and operation conditions.

<u>4.1.1 Degradation by DLD formation</u> The degradation of GaAlAs/GaAs lasers observed in the early stage of aging is usually initiated by the formation of DLDs. It was confirmed by EBIC or photoluminescence observations of degraded lasers (Deloach et al., 1973) that dark lines crossed the laser stripes. Such dark lines were light absorbing as well as nonradiative, thus increased the threshold of laser action.

The DLDs are dislocation networks staying mostly in the laser active layer. They develop typically in ⟨100⟩ direction and sometimes in ⟨210⟩ and ⟨110⟩ directions. Such dislocation networks originate from threadening dislocations in the active layer or the dislocations introduced during crystal growth. In fact, the reduction of the dislocation density in GaAs substrate crystals reduced the occurrence of the DLD formation or rapid degradation substantially (Petroff and Hartman, 1973).

4.1.2 Mechanisms of DLD growth There are two microscopic models which explain the growth of the dislocation networks: climb and glide. In the climb model, point defects such as interstitial Ga atoms are supposed to move to the dangling bonds of dislocations. The point defect density required for the DLD growth is $\sim 6 \times 10^{17}/cm^3$ (Petroff and Hartman, 1974). The motion of point defects may be accelerated by the energy generated by nonradiative recombination of carriers at dislocations.

The glide motion in $\{111\}$ plane also explains the DLD formation in the $\langle 110 \rangle$ direction. The residual stress in laser crystals or the localized stress due to local heating by nonradiative recombination would accelerate such glide motion (Matsui et al., 1975).

4.1.3 Degradation by facet damage Semiconductor lasers degrade instantaneously when the optical output power is increased. This failure mode is called catastrophic degradation. In such a failure the facet is molten and damaged. Also DLDs are often observed near the damaged facet. The optical power flux for the local melting is 5-8 MW/cm^2.

The failure power limit P_m is a function of the pulse width P_w under pulsed operation (Imai et al., 1978) as

$$P_m \propto P_w^{-1/4} \qquad (100\ ns < P_w < 10\text{-}50\ \mu s)$$

$$P_m \sim const \qquad (P_w > 10\text{-}50\ \mu s) \qquad (4.1)$$

The mechanism of catastrophic failure is not yet clear, but it is supposedly due to the rapid nonradiative surface recombination. In GaAlAs/GaAs, the thin surface layer is light-absorbing, thus causing runaway heating.

The power density of normal aging is typically 1-2 MW/cm^2, much less than the catastrophic failure limit. Even in such a case, laser facets are eroded after long term aging. The facet oxidation takes place substantially only in the illuminated portion of the laser facet, so that photo-chemical reaction may occur to promote facet oxidation (Suzuki and Ogawa, 1977).

A simple model which explains the effect of facet oxidation on laser reliability is based on the fact that the oxide film growth increases the transmissivity of the mirrors, thereby increasing the threshold current (Yuasa et al., 1978).

As discussed in Chapter 1, the threshold current I_{th} is a function of the mirror loss as

$$I_{th} \propto \alpha_m + \text{const.} \tag{4.2}$$

where

$$\alpha_m = \ln(1/r_1 r_2)/L \tag{4.3}$$

The mirror loss is maximized for an oxide film thickness of 0.12 μm.

Such facet erosion and related laser degradation have been clearly demonstrated either by high power aging or by high temperature high humidity aging. In order to prevent the facet oxidation, laser facets are coated by dielectric films such as

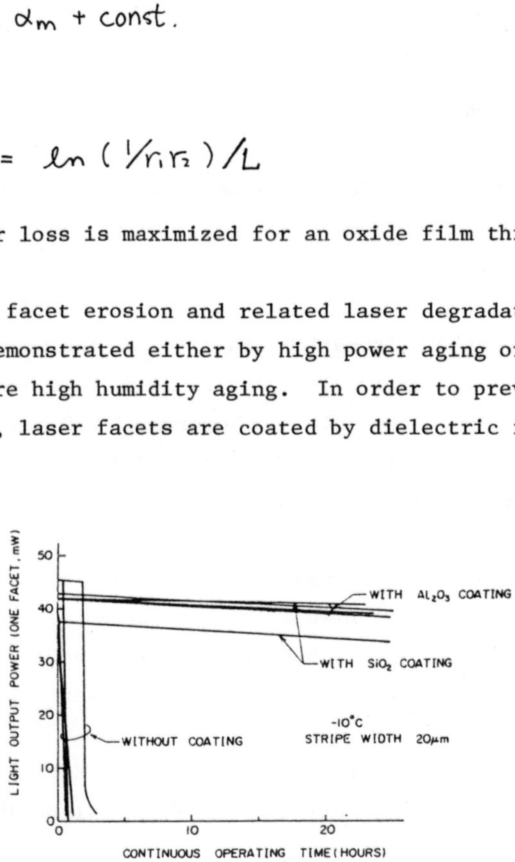

Fig. 4.1 Aging results of GaAlAs/GaAs lasers with and without facet coating (Shima et al., 1977).

Al_2O_3, Si_3N_4, and SiO_2. Figure 4.1 compares the reliability of GaAlAs/GaAs lasers with and without facet coating.

4.1.4 Other degradation modes During long term aging of GaAlAs/GaAs lasers, slow degradation takes place without appreciable DLD growth or facet erosion. The luminescence efficiency is gradually degraded. Increase of nonradiative point defects are supposed to be the origin of the uniform degradation.

4.2 Degradation Mechanisms of InGaAsP/InP Lasers

The reliability of InGaAsP/InP lasers is under study. The results obtained up to now show clear contrast to the degradation of GaAlAs/GaAs lasers.

4.2.1 Degradation by DLD formation In spite of the many threadening dislocations in the active layer of InGaAsP, DLDs are hardly observed to grow. There are only few reports of DLD observations (Yamamoto et al., 1978). The growth rate of DLDs is much slower in InGaAsP/InP lasers than in GaAlAs/GaAs lasers.

4.2.2 Degradation by facet damage The catastrophic failure limit of InGaAsP/InP lasers was studied by high power pulsed operation of BH lasers grown on a p-InP substrate (Nakano, 1981). Catastrophic failure was not observed to an optical power level of $\sim 100\ MW/cm^2$.

Under normal aging at a few MW/cm^2, facet oxidation is not noticeable after 10^4 hour operation (Mizuishi et al., 1980).

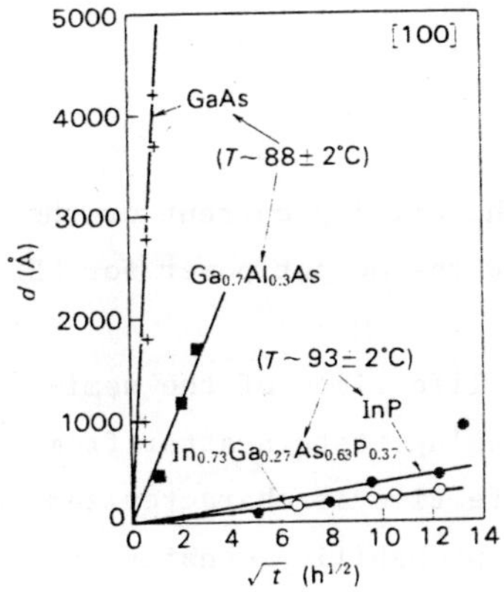

Fig. 4.2 Thickness of oxidation products of InP, InGaAsP, GaAs, and GaAlAs (Morimoto et al., 1980).

The stability of the laser facet was confirmed by immersing laser diodes into water. Figure 4.2 shows the thickness of the oxide films on InP and InGaAsP surfaces as functions of immersion time. The data on GaAlAs and GaAs are also shown for comparison. The thickness of each oxide film increased by the square root of the immersion time. The slope for InP and InGaAsP were about two orders less than that of GaAs, indicating the stability of the facets of InGaAsP/InP lasers.

4.3 Life Time Evaluation

<u>4.3.1 Accelerated aging</u> The life time or the degradation rate of semiconductor light sources is estimated by accelerated aging. There are several accelerating factors such as output power, driving current, and temperature. Among them, the temperature acceleration is a widely used method, where one assumes that the degradation rate is described by Arrhenius relation

$$dY/dT = \tilde{a} \exp(-E_A/kT_j) \tag{4.4}$$

In this relation,

$\tilde{a}$: constant

E_A : activation energy

T_j : junction temperature

and the physical parameter Y may be the driving current or the threshold current for laser diodes and the output power for LEDs.

<u>4.3.2 Weibull plot of life time</u> The life times of the semiconductor light sources subjected to aging tests scatter from sample to sample. Such scatter of life time is characterized by Weibull distribution, where a failure probability density function is

$$f(t) = \begin{cases} \frac{m}{\alpha}(t-\gamma)^{m-1} \exp\left[-\frac{(t-\gamma)^m}{\alpha}\right] & (t \geq \gamma) \\ 0 & (t < \gamma) \end{cases} \quad (4.5)$$

The failure modes are categorized in three groups:

(1) $m < 1$ infant mortality,

(2) $m = 1$ random failure, and

(3) $m > 1$ wear-out

Using Equation 4.5, one gets

$$\text{MTTF} : \quad \mu = \alpha^{\frac{1}{m}} \Gamma\left(\frac{1}{m} + 1\right) \quad (4.6)$$

In these equations,

γ : location parameter

α : scale parameter

m : shape parameter

4.3.3 Log-normal plot of life time Another distribution which is often used is the log-normal distribution. It is given by the following relations:

Failure probability density function

$$f(t) = \frac{\log_{10} e}{\sqrt{2\pi}\,\sigma t} \exp\left[- \frac{(\log t - \log \xi)^2}{2\sigma^2} \right] \tag{4.7}$$

MTTF;

$$\mu = \xi \exp(\sigma^2/2) \tag{4.8}$$

where

σ : spread factor

ξ : median life

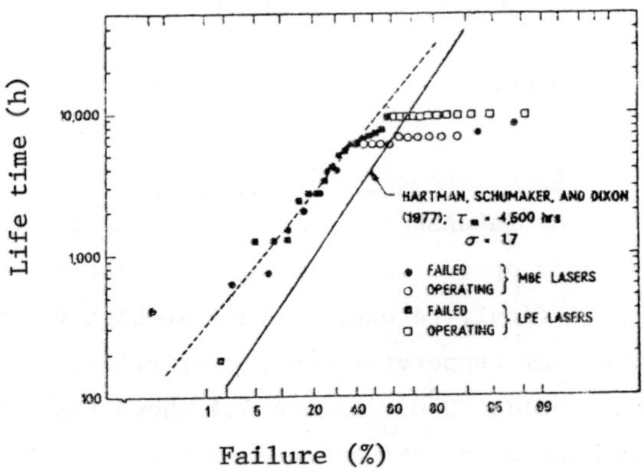

Fig. 4.3 Life time distribution of GaAsAs/GaAs lasers fabricated by LPE and MBE (Tsang, et al., 1981).

4.3.4 Reliability of GaAlAs/GaAs lasers In the evaluation of the life time of GaAlAs/GaAs lasers, it is usual to prescreen them in order to eliminate the infant mortality caused by DLD growth etc. The laser screening is done at 50-70°C for 20-100 hours. Figure 4.3 shows an example of the log-normal plot of degradation rate, where lasers were aged at 70°C. There are some deviation of data from the straight line, especially for short life lasers. But in general, they are well approximated by the log-normal distribution. A median life of 8000 hours was obtained at 70°C. The activation energy reported on GaAlAs/GaAs lasers are ranging from 0.7-0.9 eV. Using these values, one gets a room temperature extrapolated life time of more than 10^6 hours.

The reliability of GaAsAl/GaAs lasers does not depend strongly on laser wavelength for λ=0.78-0.86 µm. Below 0.78 µm, poor reliability was reported (Kajimura et al. 1980).

The laser reliability was also shown to depend on laser current I_0. Mizuishi et al. (1979) assumed the following relation to describe the laser degradation rate

$$dI/dt \propto \exp(-E_A/kT_j) \cdot [\exp(\tilde{p} I_0) - 1]$$

where $\tilde{p}$ is a constant.

4.3.5 Reliability of InGaAsP/InP lasers The recent results on the reliability of InGaAsP/InP lasers indicate the capability of long-term reliable operation at 1.3 µm. In the case of InGaAsP/InP lasers, it is difficult to operate them at high temperatures because of the poor temperature characteristics. Usually lasers are aged at and below 70°C. Figure 4.4 shows the aging of InGaAsP/InP BH lasers, where an activation energy of 0.6 eV was obtained. From the median time of 1.7×10^4 h at 70°C, one estimates the room temperature life time to be $> 1 \times 10^6$ hours (Mizuishi, 1982).

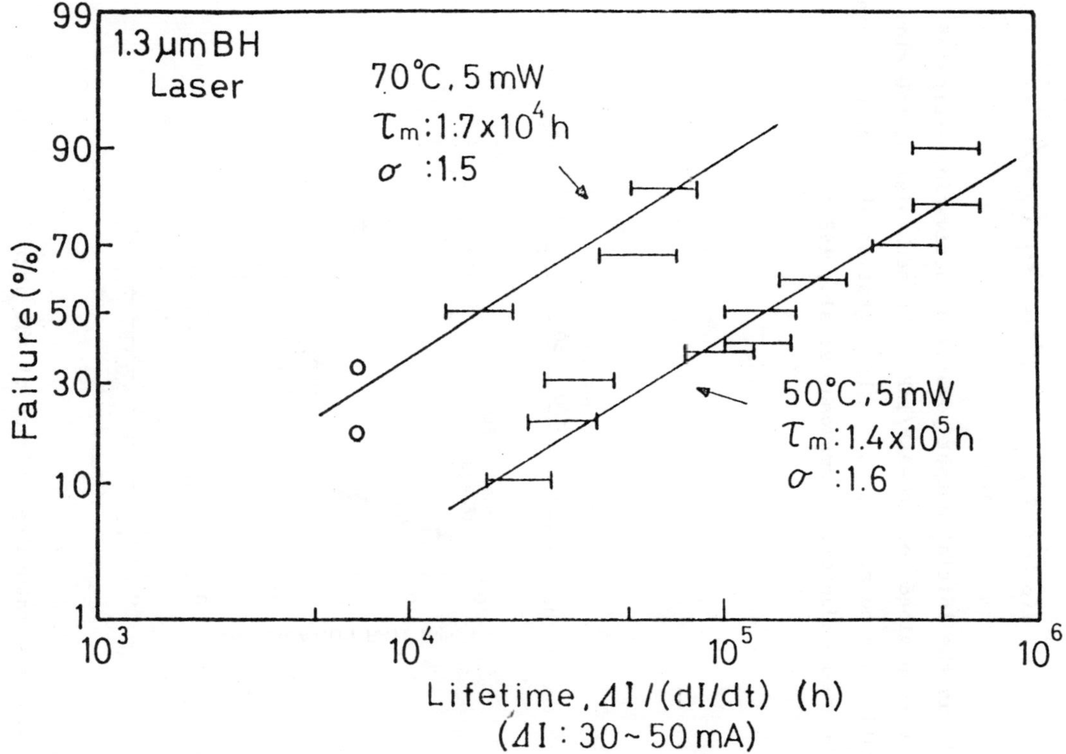

Fig. 4.4 Temperature dependence of degradation rate of InGaAsP/InP lasers at 1.3 μm. (Mizuishi et al., 1982)

4.3.6 Reliability of LEDs The degradation rate of InGaAsP/InP and GaAsAl/GaAs LEDs are shown in Figure 4.5, where the degradation rate was defined by

$$P = P_o \exp(-\beta^* t), \qquad \beta^* \propto \exp(-E_A/kT_j)$$

where P_0 is the initial output power. An activation energy of 1.0 eV was obtained for InGaAsP/InP. The mechanism of the slow degradation was considered to be the migration of electrode metal into the semiconductor. (Yamakoshi et al., 1981).

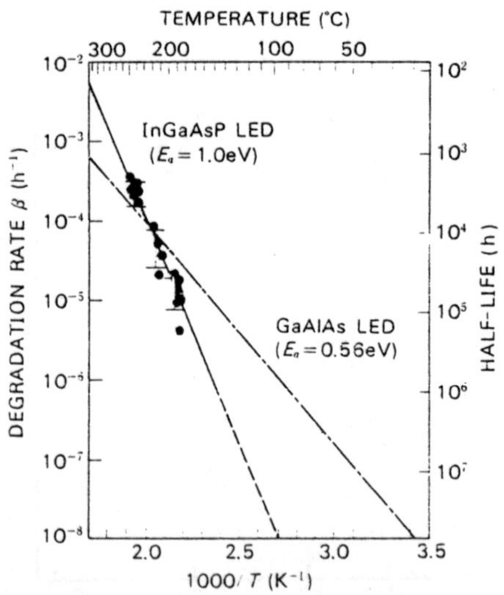

Fig. 4.5 Temperature dependence of the degradation rate of InGaAsP/InP and GaAlAs/GaAs LEDs (Yamakoshi et al. 1981).

REFERENCES

DeLoach, B.C., Jr., et al., 1973, Proc. IEEE 61, 1042.

Imai, H., et al., 1978, Appl. Phys. Lett. 33, 1011.

Kajimura, T., 1980, J. Appl. Phys. 51, 908.

Matsui, J., Ishida, K., and Nannichi, Y., 1975, Japan J. Appl. Phys., 14, 1555.

Mizuishi, K., Chinone, N., Sata, H., and Ito, R., 1979, J. Appl. Phys. 50, 6608.

Mizuishi, K., et al., 1980, Japan. J. Appl. Phys. 19, 1429.

Mizuishi, K., 1982, private communication.

Morimoto, M., et al., 1980, Appl. Phys. Lett. 37, 1082.

Nakano, Y., private communication.

Petroff, P.M., and Hartman, R.L., 1973, Appl. Phys. Lett. 23, 461.

Petroff, P.M., and Hartman, R.L., 1974, J. Appl. Phys. 45, 3899.

Shima, N., et al., 1977, Appl. Phys. Lett. 31, 625

Suzuki, T., and Ogawa, M., 1977, Appl. Phys. Lett. 31, 473.

Tsang, et al., 1981, Appl. Phys. Lett. 39, 683.

Yamakoshi, S., et al., 1981, IEEE J. Quantum Electron., QE-17, 167.

Yamamoto, T., Sakai, K., and Akiba, S., 1978, IEEE J. Quantum Electron. QE-15, 684.

Yuasa, T., et al., 1978, Appl. Phys. Lett. 32, 119.

5. PHOTODETECTORS

The requirements for semiconductor photodetectors are (1) high efficiency, (2) low noise, and (3) high speed operation. To meet them, Si PIN photodiodes and APDs have been developed for $\lambda \sim 0.8\ \mu m$. Ge and III-V compound materials are used for 1.1-1.6 μm.

5.1 Basic Characteristics of Photodiodes (Anderson et al., 1966)

The basic concept of a pn photodiode is shown in Figure 5.1. The electrons and holes generated by light illumination are accelerated by the electric field. The resulting current is detected by the external load.

The photocurrent I_p is assumed to be proportional to the number of incident photons:

$$I_p = \eta \left(\frac{qP}{h\nu} \right) M \qquad (5.1)$$

where P is the incident optical power and η is the quantum efficiency. Associated with the average photocurrent is a mean square shot noise:

$$\overline{i_s^2} = 2\, q I_p B M^2 F \qquad (5.2)$$

where B is the band width.

In these equations, M is the multiplication factor denoting avalanche multiplication of carriers in the depletion region. In PIN photodiodes, M = 1. The factor F is the excess noise factor mentioned later. The load also generates thermal noise which is given by

$$\overline{i_T^2} = 4kT_eB/R_L \qquad (5.3)$$

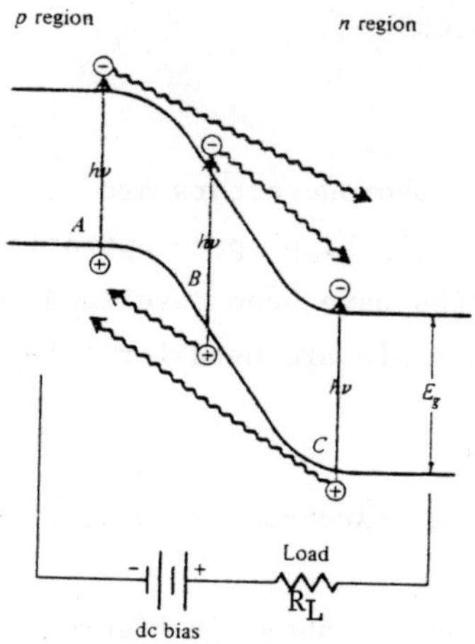

Fig. 5.1 Concept of a pn photodiode.

where T_e is the effective noise temperature. Taking these into account, the equivalent circuit of a photodiode is given in Figure 5.2. Then the signal to noise ratio S/N is given by

$$S/N = \frac{\frac{1}{2}\left(\frac{\eta q P}{h\nu}\right)^2 M^2}{\left[\frac{2\eta q}{h\nu}(P+P_B)+2qI_D\right]BM^2F + \frac{4kT_eB}{R_L}} \tag{5.4}$$

where

P, P_B : incident and background powers

B : band width

I_D : dark current

In this equation, the dark current is also included, which is the sum of the junction dark current and the surface leakage current.

5.1.1 Excess noise factor The multiplication noise in APDs originates from the random process of multiplication in the avalanche region. It becomes large when the electron ionization rate α and the hole ionization rate β are similar to each other. When only electrons or holes trigger the avalanche multiplication, the excess noise factors are given by (McIntyre, 1966):

$$F_n = M_n [1 - (1-k)(1 - 1/M_n)^2]$$

$$F_p = M_p [1 - (1-K^{-1})(1 - 1/M_p)^2] \qquad (5.5)$$

$$K = \beta/\alpha$$

In order to realize low avalanche noise APDs, the k value should be much different from unity. The normalized excess noise factor F/M is plotted in Figure 5.3. For $\alpha = \beta$, the ratio k=1, and F=M.

5.1.2 Minimum detectable power in PIN photodiodes The thermal noise is usually larger than the shot noise. Thus the S/N ratio is

$$S/N \simeq (\eta P q/h\nu)^2 \Big/ \frac{8kT_eB}{R_L} \qquad (5.6)$$

The minimum detectable power is defined as the power at which S/N=1, and is given by

$$P_{min} = \frac{h\nu}{q\eta}\sqrt{\frac{8kT_eB}{R_L}} \qquad (5.7)$$

5.1.3 Minimum detectable power in APDs As M increases, the signal and the shot noise term in Equation 5.4 increase. The resultant S/N increases with M until the shot noise term becomes comparable with the thermal noise term. The minimum detectable power is then given by

$$P_{min} = \frac{h\nu}{M'q\eta}\sqrt{\frac{8kT_eB}{R_L}} \qquad (5.8)$$

where M' is the multiplication factor at which the shot noise is equal to the thermal noise. The optimum M' is 30-200 in Si APDs; thus a considerable improvement is achieved in sensitivity over that of PIN photodiodes.

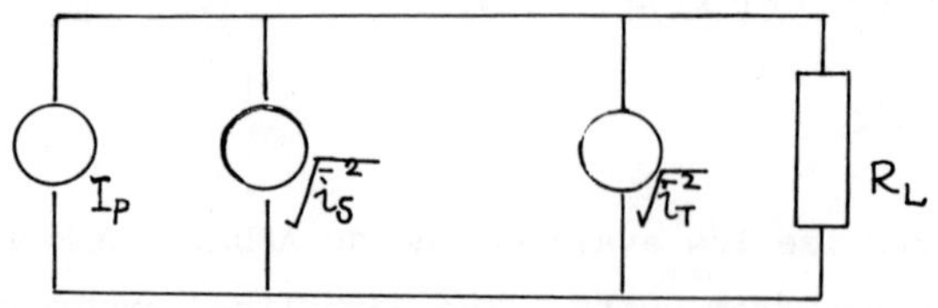

Fig. 5.2 Equivalent circuit of a photodetection circuit.

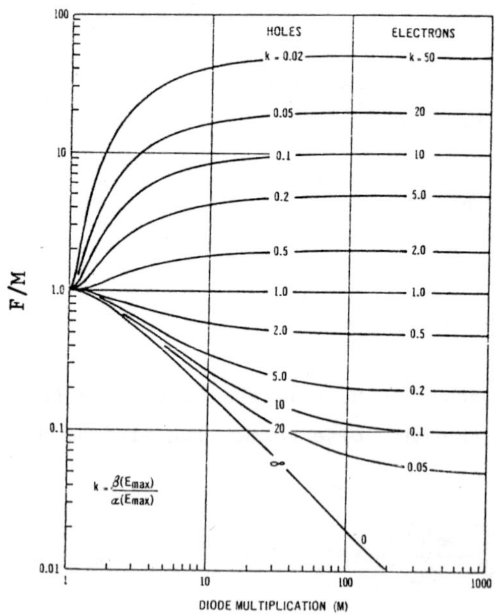

Fig. 5.3 Normalized excess noise factor F/M (McIntyre, 1966).

5.1.4 Frequency response The frequency response of semiconductor photodetectors is determined by the shunting effect of the signal current by the junction capacitance C_j:

$$\omega_m = \frac{1}{(R_i + R_L) C_j} \tag{5.9}$$

where R_i is the diode resistance and ω_m is the cut-off angular frequency. For high speed detection of light, the junction capacitance should be small.

5.2 Si Photodetectors

5.2.1 PIN photodiodes PIN photodiodes have the advantage of low operating voltage. In order to realize a low voltage device, the i layer must have as low an impurity concentration as possible. An i layer with an impurity concentration of $5 \times 10^{12}/cm^3$ is obtained by the high purity epitaxial growth technique. Figure 5.4 shows the bias voltage dependence of typical PIN photodiode of Si. The cut-off frequency of 3 dB down is 340 MHz and 1 GHz at bias voltages of 5 V and 30 V respectively. The external quantum efficiency is about 65% at 0.83 μm. The junction capacitance is 0.8 pF at 5 V bias condition for the diode with an active area of 300 μm in diameter.

5.2.2 APDs Figure 5.5 shows an example of Si APDs (Ohuchi, 1981). This structure has an extremely thin p-layer between the p^- and π layers, based on theoretical consideration. The p layer serves in reducing background voltage without deteriorating any other performances. The trade-off observed between noise and operating voltage can be remarkably improved by optimizing the impurity concentration and the width of the p-side regions. The APD is designed to have a p^- layer with an acceptor concentration as low as possible. In this layer the avalanche multiplication takes place. The n^+-p^--p-π-p^+ has a p layer 3.5 μm wide and acceptor concentration of $2x10^{15}/cm^3$, and a π layer with $1\text{-}5x10^{13}/cm^3$ acceptor concentration.

The dependence of excess noise factor on multiplication factor was examined and the effective k ratio was evaluated to be 0.03, showing good agreement with the theoretical curve. The other parameters of the APDs are as follows: a gain band width parameter of 250 MHz at multiplication factor of 500, a breakdown voltage about 140 V, and an active area of 300 μm in diameter.

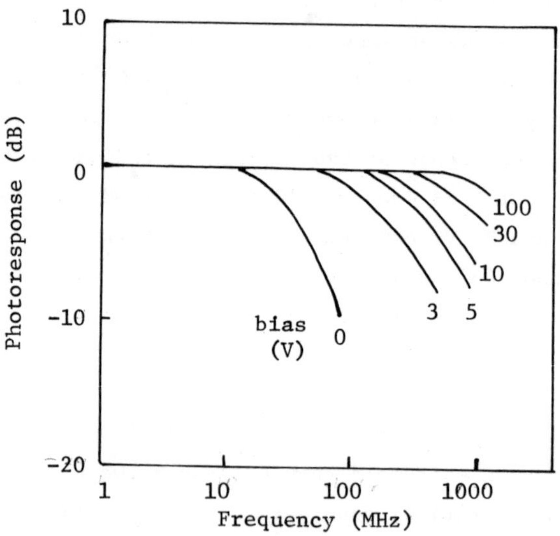

Fig. 5.4 Bias voltage dependence of a typical Si PIN photodiode.

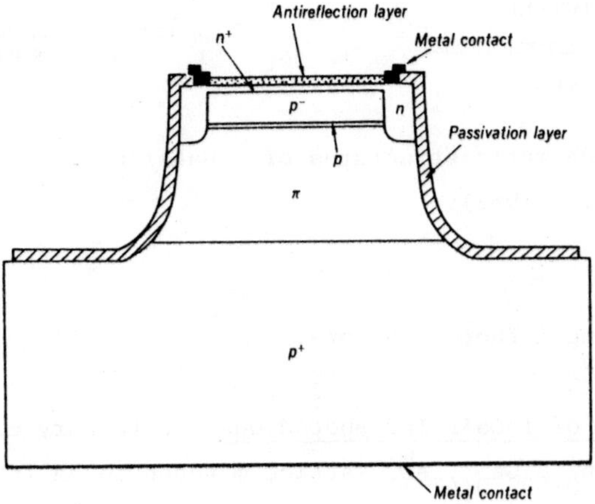

Fig. 5.5 Schematic structure of a Si APD for low excess noise and low operating current.

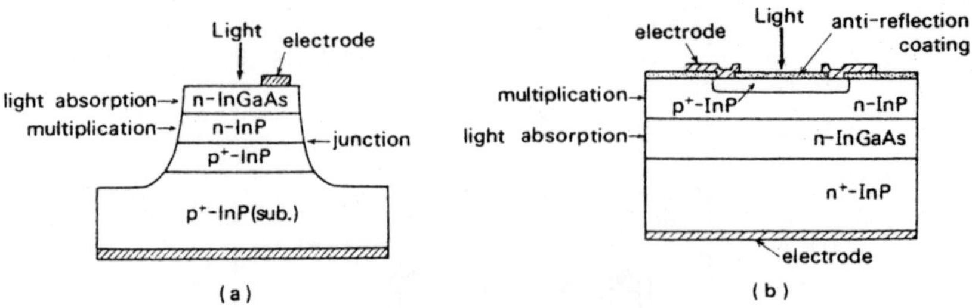

Fig. 5.6 Schematic structures of InGaAs/InP photo-detectors (Kanbe et al., 1982).

5.3 Long Wavelength Photodetectors

<u>5.3.1 Structure of InGaAs/InP photodiodes</u> A ternary compound semiconductor, $In_{0.53}Ga_{0.47}As$, lattice matched to an InP substrate is a suitable material for long wavelength photodetection. A typical structure employed presently is shown in Figure 5.6.

Because of the small bandgap energy and the small effective electron mass of InGaAs, the p-n junction fabricated in InGaAs had large tunneling current which would be given by

$$J_t = \sqrt{\frac{2m}{E_g}}\,(q^3 E w_t / 2\pi h)\exp\left[-\pi^2 (2m_e E_g^3)^{1/2} / 2qhE\right] \qquad (5.10)$$

where E is the electric field strength, w_t is the tunneling region thickness, and m_e is the effective electron mass. Figure 5.7 shows the voltage for Zener tunneling calculated by the above expression together with the avalanche breakdown voltage (Kanbe et al., 1982). The avalanche breakdown voltage is approximately given by

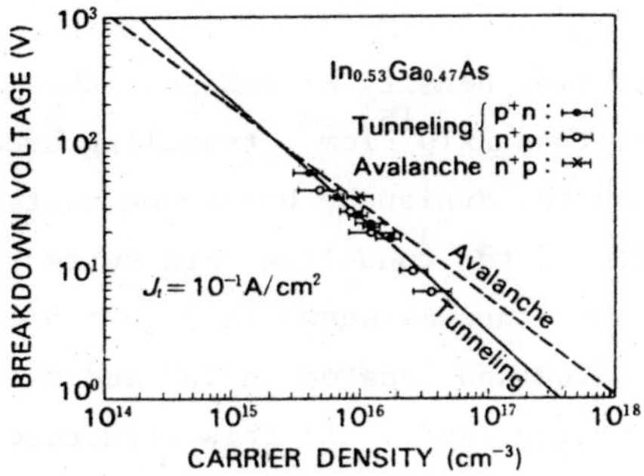

Fig. 5.7 Comparison of Zener tunneling breakdown voltage and avalanche breakdown voltage (Kanbe et al., 1982).

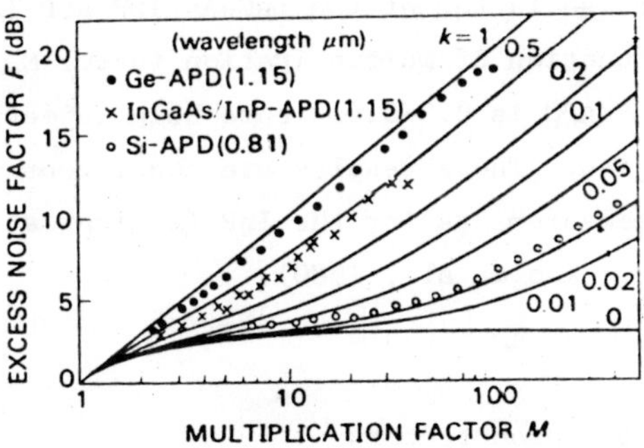

Fig. 5.8 Excess noise factors for Ge, InGaAs, and Si (Kanbe et al., 1982).

$$V_B \doteq 60\,(E_g/1.1)^{3/2}\,(\bar{n}/10^6)^{-3/2} \tag{5.11}$$

where n is the carrier density of InGaAs. When the carrier density is higher than $2\times10^{15}/cm^3$, tunneling breakdown voltages becomes lower than the avalanche breakdown voltage.

The reduction of the tunneling dark current was the motive force to have a structure as shown in Figure 5.6, where a p-n junction is separated and located in InP and separated from the InGaAs light absorbing layer. In this structure, the electric field is low in the InGaAs layer, thus the tunneling current should be much less.

<u>5.3.2 Characteristics of InGaAs/InP photodetectors</u> (Kanbe et al., 1982). The quantum efficiency of the diodes were about 52% at a wavelength of 1.5 μm and about 40% at 1.3 μm without AR coating. The dark current was in the order of 10^{-4} A/cm^2 at $0.95V_B$.

The excess noise factor of the InGaAs/InP APD is shown in Figure 5.8 as a function of multiplication factor M. The k value for the InGaAs/InP APD is 0.3-0.4. Pure hole injection occurs in the avalanche region. These results are consistent with the impact ionization rate measurements for the InP (Armient et al., 1979; Umebu et al., 1980; Kao et al., 1980).

5.4 New Structure Photodetectors

The drawback of using III-V compound semiconductor materials such as GaAs is the fact that the ratio of ionization factors, k, is not so much different from unity. To overcome this problem, new photodetector structures have also been proposed.

Among them is a photodetector consisting of a graded $Ga_{1-x}Al_xAs$ layer and a $Ga_{0.55}Al_{0.45}As$ layer grown by MBE (Cappaso et al., 1981), where the generated electrons are accelerated towards the narrow bandgap region, while holes come to the wide band gap region. Thus the electron ionization factor is higher than the hole ionization factor. In fact the ratio k of 10 was obtained in such a graded structure.

REFERENCES

Anderson, L.K., and McMurtry, B.J., 1966, Proc. IEEE 54, 1335.

Armiento, G.A., et al., 1979, Appl. Phys. Lett. 35, 333.

Capasso, et al., 1981, presented at the 1981 Symposium on GaAs and Related Compound Materials.

Kanbe, H., Susa, N., and Ando, H., 1982, in Optical Devices and Fibers, edited by Y. Suematsu (OHM, Tokyo).

Kao, C.W., and Crowell, C.R., 1980, Solid State Electron. 23, 881.

McIntyre, R.J., 1966, IEEE Trans. Electron Devices ED-13, 164.

Ohuchi, H., 1981, private communication.

Umebu, I., et al., 1980, Appl. Phys. Lett. 36, 302.

6. CRYSTAL GROWTH AND PROCESS OF OPTOELECTRONIC MATERIALS

Along with the development of device structures and characterization, the fabrication technology of III-V compound semiconductor devices have amazingly been improved. In optoelectronic devices, crystal growth is the key issue since the device performance is greatly determined by the quality of the crystal itself. Liquid phase epitaxy, molecular beam epitaxy and chemical vapor deposition are the crystal growth methods employed presently. Also important are electrode formation, passivation by dielectric films, and crystal microfabrication. The up-dated fabrication technologies are summarized in this chapter.

6.1 Liquid Phase Epitaxy

<u>6.1.1 Phase diagram of GaAlAs</u> Assuming that a regular ternary liquid solution is in equilibrium at temperature T with an ideal solid solution of GaAs and AlAs (Illgems et al., 1968), the equations describing the isotherms are given by;

$$1-x = 4N_{Ga}N_{As}\frac{\gamma_{Ga}\gamma_{As}}{\gamma^{sl}_{Ga}\gamma^{sl}_{As}}\exp\left\{\Delta S^F_{GaAs}(T^F_{GaAs}-T)/kT\right\}$$

$$x = 4N_{Al}N_{As}\frac{\gamma_{Al}\gamma_{As}}{\gamma^{sl}_{Al}\gamma^{sl}_{As}}\exp\left\{\Delta S^F_{AlAs}(T^F_{Al}-T)/kT\right\} \qquad (6.1)$$

$$N_{Ga}+N_{Al}+N_{As}=1$$

where

γ_i : activity coefficient in liquid

N_i : atom concentration in liquid

x : mole fraction of AlAs in solid

ΔS^F_i : entropy of fusion

T^F_i : temperature of fusion

γ^{sl}_i : activity coefficient of binary stoichiometric liquid

The activity coefficients are given by the following equation and cyclic permutation of the indices

$$RT \log \gamma_{Ga} = \alpha_{GaAs} N_{As}^2 + \alpha_{GaAl} N_{Al}^2 + (\alpha_{GaAs} - \alpha_{AlAs} + \alpha_{GaAl}) N_{As} N_{Al}$$

where α_i are the interaction parameters of the indicated system. In Table 6.1 are listed the parameters necessary for the calculation of isotherms. Figure 6.1 shows an example of GaAlAs solidus compositions as a function of liquidus composition (Panish and Hayashi, 1974). The agreements between the experimental and theoretical results are good.

6.1.2 Phase diagram of InGaAsP The phase diagram of InGaAsP is also calculated assuming a regular quaternary liquid solution in equilibrium with an ideal solid solution of GaAs, InAs, GaP, and InP. The equations describing isotherms are similar to Equation 6.1 (Antypas and Edgecumbe, 1976).

Table 6.1 Parameters for calculating phase diagrams.

	T_F(K)	S_F (e.u.)	α (cal/mole)
Ga-As	1511	16.64	-9.16T + 5160
Al-As	2043	15.6	-5.5 T + 6390
Ga-P	1740	16.3	-7.0 T + 7900
In-As	1215	14.52	-9.16T + 4300
In-P	1335	14.7	0
Ga-In	-	-	1066
As-P	-	-	2000
Ga-Al	-	-	104

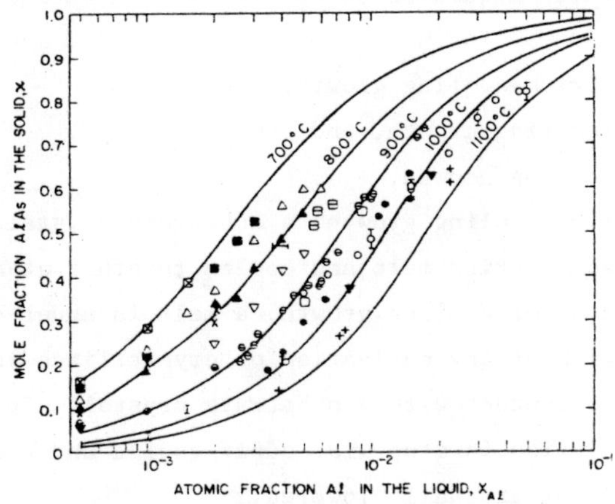

Fig. 6.1 Solidus compositions in $Ga_{1-x}Al_xAs$ as a function of liquidus composition (Panish and Hayashi, 1974).

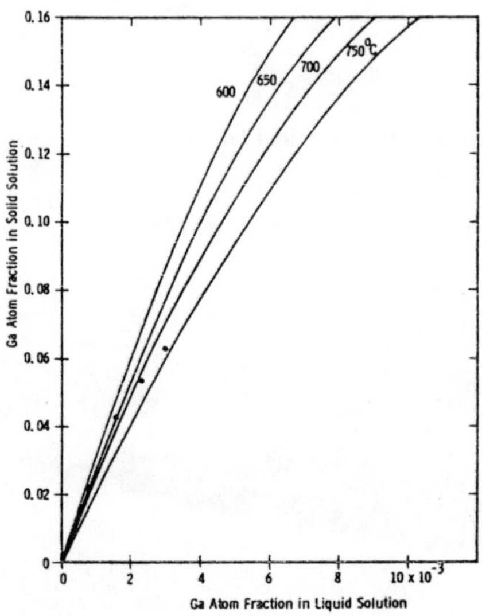

Fig. 6.2 Solidus composition in InGaAsP. $X^l_{As} = 5.65x10^{-3}$ (Antypas and Edgecumbe, 1976).

6.1.3 Growth rate Liquid phase epitaxy may be categorized in three types

(1) equilibrium-cooling growth,

(2) super-cooling growth, and

(3) step-cooling growth.

In the equilibrium cooling growth, a substrate crystal is contacted at T_1 with an equilibrium melt and cooled together at a cooling rate a. In the super cooling growth, a melt is super-cooled by an amount of ΔT without any nucleation of crystallites in the melt, and brought into contact with a substrate crystal. In the step cooling, the solution is step-wise super-cooled and the temperature is kept constant or gradually lowered.

The thickness of GaAlAs grown by these methods are well approximated by solving a diffusion equation of As in the Ga solution and given by

$$d = \frac{1}{C_{As}^{sl}\, m}\left(\frac{D}{a}\right)^{1/2}\left[2\Delta T\, t^{1/2} + \frac{4}{3} a t^{3/2}\right] \tag{6.2}$$

where

C_{As}^{sl} : As concentration in solid

m : slope of liquids curve

D : diffusivity of As in the solution.

Fig. 6.3 A graphite boat used for liquid phase epitaxy.

6.1.4 Lattice matching At growth temperature of $\sim 800^{\circ}C$, the lattice constant of GaAlAs is almost the same as that of GaAs. Upon cooling, considerable stress is introduced because of the difference in the expansion coefficients (Rozgonyi et al., 1974).

For more perfect lattice matching, GaAlAsP was studied (Rozgonyi et al., 1973). By adding a small amount of P, the lattice constant of the epitaxial layer can be matched to that of GaAs at room temperature. The P addition was expected to improve the laser reliability, but no conclusive results have been obtained.

The lattice constant of InGaAsP can be matched to that of InP either at room temperature or at the growth temperature. Lattice mismatching is usually less than 0.03% in conventional liquid phase epitaxy.

In practical growth of InGaAsP, it is important not to carry over the melt to the next solution chamber. Otherwise, misfit dislocations would be generated at the hetero-interface, causing degradations in electrical and optical characteristics.

6.2 Molecular Beam Epitaxy (MBE)

Molecular beam epitaxy takes place in ultra high vacuum by impinging molecular and atomic beams to a heated substrate crystal. Since the successful application of MBE to GaAs, it has widely been applied to various compound crystals. The characteristic features of the MBE are (Ploog, 1980):

(1) a low growth rate (1-3 Å/s for GaAs),

(2) a low growth temperature ($\sim 630^{\circ}C$ for GaAs),

(3) abrupt initiation and cessation of growth,

(4) smooth growth surface, and

(5) the ability of in situ analysis.

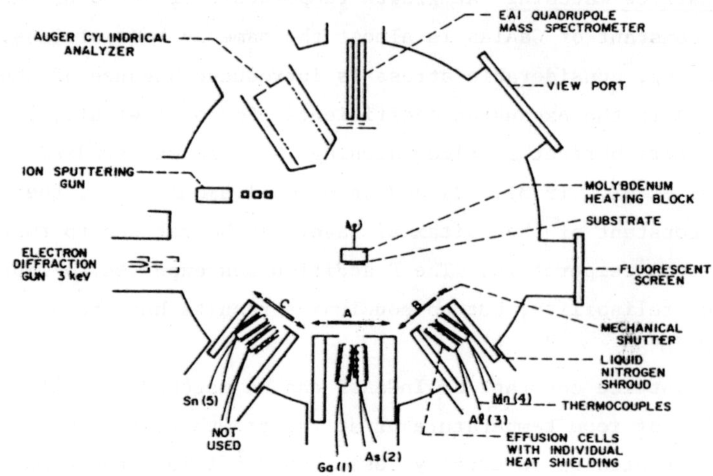

Fig. 6.4 Schematic presentation of an MBE system for the growth of GaAlAs (Cho and Arthur, 1975).

Figure 6.4 shows a schematic presentation of an MBE apparatus, consisting of a high vacuum chamber, BN fusion cells, and a substrate crystal holder. The pumping system is capable of evacuation of the system to 10^{-10}-10^{-11} torr before crystal growth.

The growth rate R of MBE GaAlAs is given by

$$R \approx \gamma_{Ga} F_{Ga} + \gamma_{Al} F_{Al} \tag{6.3}$$

where γ_{Ga} and γ_{Al} are the sticking coefficients of Ga and Al, F_{Ga} and F_{Al} are the fluxes of Ga and Al at the substrate. The sticking coefficients of Ga and Al atoms are very close to unity at typical growth temperatures under As stabilized conditions. The mole fraction of AlAs in the GaAlAs is thus the ratio of the Al flux to the total flux of Ga and Al.

The flux of atoms or moleculs of species i at the substrate is given by

$$F_i = \frac{SP_i}{\pi d^2\sqrt{2\pi m_i kT}} \cos\theta \qquad (\text{molecules}/\text{cm}^2\cdot\text{s}) \qquad (6.4)$$

where

S : area of the cell aperture,

P_i : equilibrium vapor pressure,

d : distance of the orifice from the substrate,

m_i : mass of the effusing species,

θ : angle between the beam and the normal of the substrate.

For typical growth rate of 1 μm/h, the flux is 10^{15}-10^{16}/cm^2.s for the group V elements, 10^{14}-10^{15}/cm^2.s for group III elements, and 10^8-10^9/cm^2.s for the dopants.

6.3 Metal-Organic Chemical Vapor Deposition (MO-CVD)

MO-CVD is a growth technique developed to grow thin uniform layers of III-V compound crystals. Crystal growth takes place as a result of the thermal decomposition of both the metal-organic compounds and hydride, in such a way as

$$Ga(CH_3)_3 + AsH_3 \xrightarrow{H_2} GaAs + 3CH_4$$

Doping is generally done by introducing organometallics ($Zn(C_2H_5)_2$ etc.) or hydrides (H_2S etc.) into the reaction furnace. A schematic presentation of MO-CVD apparatus is shown in Figure 6.5 (Dupui and Dapkus, 1977).

In the case of the quaternary material (InGaAsP), the growth chamber is low-pressured to avoid the direct reaction of PH_3 with In.

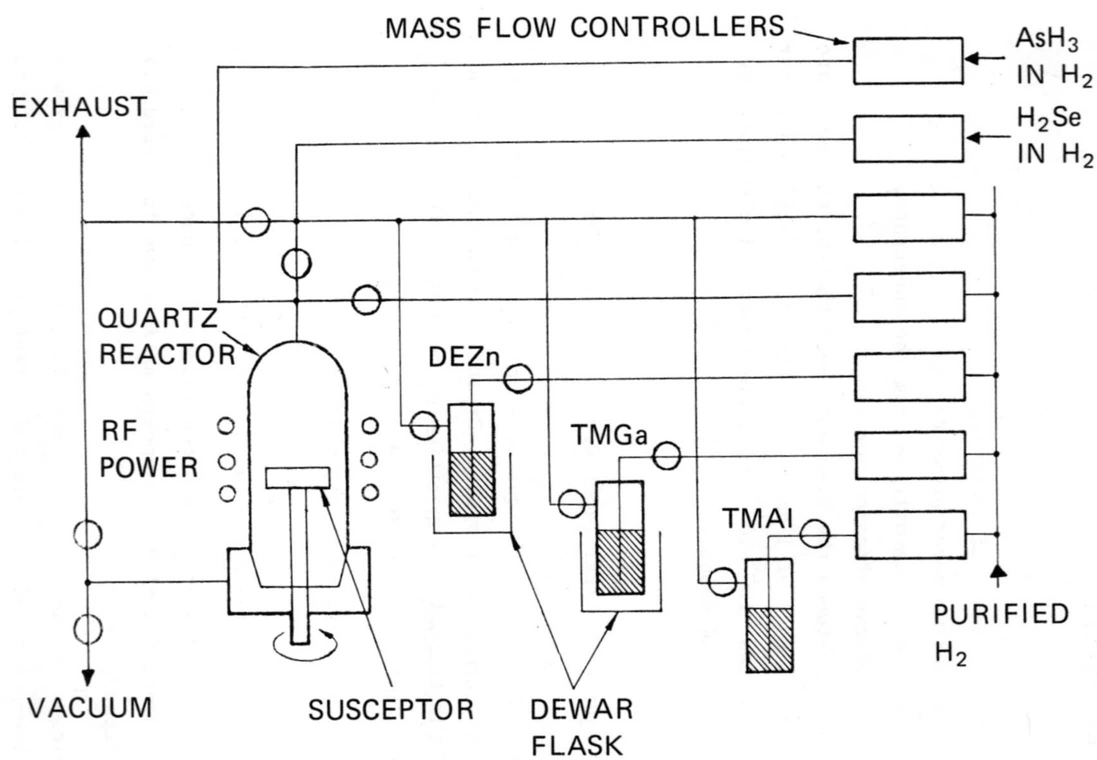

Fig. 6.5 Schematic presentation of MO-CVD apparatus for GaAlAs (Dupui and Dapkus, 1977).

6.5 Metallization

Various types of electrode metals have been used to fabricate ohmic contacts to GaAs and InP. The mechanism of current flow is explained by the tunneling model. In order to fabricate low contact resistance ohmic contacts, the surface carrier density is the most crucial parameter. High carrier density is obtained either by impurity diffusion or alloying.

The contact resistance is typically $10^{-6}\ \Omega.cm^2$ for n-GaAs and n-InP under optimized conditions, and 10^{-4}-$10^{-5}\ \Omega.cm^2$ for p-GaAs and p-InP under optimized conditions.

The use of refractory metals are of recent concern in order to prevent Au diffusion into crystal, which has been found to degrade the reliability of LEDs and lasers.

6.6 Passivation by Dielectric Films

As mentioned in Chaper 4, facet coating of laser mirrors is necessary for long-term operation of GaAlAs lasers. In the coating process, the cleanliness of the initial facet was found to be important, therefore the facet coating is done immediately after the cleavage or after cleaning by sputtering or some other means.

6.7 Mirror Formation

Cleavage is the only practical technique of facet formation. It gives perfect mirrors composed of (110) crystalline facets. However, replacing it by an etching method has also been studied by many researchers.

Facet formation by etching is desirable because the following processes such as facet coating and testing could be possible without making individual chips. Also for the integration of lasers with other optical and electrical devices, cleaved facets should be replaced by etched facets or something else.

If the laser stripe is $\langle 100 \rangle$ oriented, crystalline facets of $\{100\}$ are likely to appear by wet etching. In practice, however, the resultant threshold current is 30-50% higher than that of cleaved lasers. It is partly because of the poor quality of the etching masks (Merz and Logan, 1976; Tarui, et al., 1976).

To fabricate vertical mirrors reproducibly, dry etching process has also been studied (Bosch et al., 1981).

REFERENCES

Antypas, G.A., and Edgecumbe, J., 1976, J. Crys. Growth 34, 132.

Bosch, M.A., Golren, L.A., and Good, E., 1981, Appl. Phys. Lett. 38, 264.

Cho, A.Y. and Arthur, J.R., 1975, in Progress in Solid-State Physics, (Pergamon Press, New York), vol. 10, p. 157.

Dupui, R.D., and Dapkus, P.D., 1977, Appl. Phys. Lett. 31, 466.

Illegems, M., and Pearson, G.L., 1969, Proc. Int. Symp. Gallium Arsenide and Related Materials, p.3.

Merz, J.L., and Logan, R.A., 1976, Appl. Phys. Lett. 47, 3503.

Panish, M.B., and Hayashi, I., 1974, in Applied Solid State Science, edited by R. Wolfe (Academic Press, New York), vol. 4, p. 235.

Ploog, K., 1980, in Crystals, edited by H.C. Freyhardt (Springer-Verlag, New York), vol. 3, p. 73.

Rozgonyi, G.A., and Panish, M.B., 1973, Appl. Phys. Lett. 23, 533.

Rozgonyi, G.A., Petroff, P.M., and Panish, M.B., 1974, J. Crystal Growth 27, 106.

Tarui, Y., Komiya, Y., Iida, H., and Shoji, A., 1976, Proc. 7th Conf. Solid State Devices, p. 264.

GHAVAM GHAVAMISHAHIDI